BUILDING THE CLIMATE CHANGE RESILIENCE OF MONGOLIA'S BLUE PEARL

A CASE STUDY OF KHUVSGUL LAKE NATIONAL PARK

Mark R. Bezuijen, Michael Russell, Robert J. Zomer, and D. Enkhtaivan

DECEMBER 2020

ASIAN DEVELOPMENT BANK

ADB

Contents

Tables and Figures

Foreword

Since the outbreak and global spread of the coronavirus disease (COVID-19), there has been renewed and growing awareness of the need to protect nature. Protected areas are fundamental in this regard: they are places that help maintain healthy populations of plants and animals and protect the natural systems upon which life is based, including ecosystem services such as water supply and flood regulation. They provide a buffer from the risk of disease outbreaks and the impacts of climate change. Maintaining protected areas in good condition is critical to help protect nature and people, yet protected areas globally are under threat. Now, more than ever, as part of efforts toward a nature-based recovery from the COVID-19 crisis, there is a need to support the design, planning, and management of protected areas.

Mongolia's protected areas encompass deserts, grasslands, and snow-capped mountains—a stunning expanse of wilderness that supports globally important biodiversity values and nomadic patterns of life largely unchanged for thousands of years. Comprising a mosaic of parks and reserves, these areas support some of the largest and most intact natural systems remaining in the world, including pristine lakes and undammed rivers.

At the same time, many of these areas suffer from limited management and funding, and support local communities that are poor, vulnerable, and have few options other than to engage in the extraction of natural resources. Nature-based tourism offers excellent opportunities to help address some of these issues by creating much-needed rural jobs, diversifying local incomes, and helping to finance park management. To secure these benefits, the government aims to expand the existing protected area network to cover almost one-third of Mongolia's total land area by 2030. Efforts are also underway to upgrade public infrastructure, facilities, and services to support nature-based tourism across the country.

Climate change threatens to undermine these nature-positive efforts. Mongolia is experiencing some of the highest rates of climate change in the world, including rising temperatures and increased frequency and intensity of drought, storms, and other climate hazards. The impacts of climate change have critical implications for the management of protected areas. How will climate change affect the role of protected areas to support nature and benefit livelihoods, and what can be done to address these risks? The answers to these complex questions will help inform policy makers and land managers, and contribute to long-term planning for conservation and development.

This study is a timely and forward-looking contribution to these issues. It examines the projected impacts of climate change on conservation, livelihoods, and tourism at Khuvsgul Lake National Park, one of Mongolia's most spectacular and visited protected areas. The park is located in the fastest-warming region of the country, northern Mongolia, where climate change is already affecting nature and people.

The Asian Development Bank (ADB) supports Mongolia's efforts to protect its biodiversity, improve livelihoods, and develop sustainable tourism. Environmental sustainability is a central pillar of ADB's country partnership strategies with Mongolia. ADB-supported efforts to help protect nature and build resilience to climate change include projects for protected area management, sustainable tourism, and disaster risk management. The current study will help to inform project designs and strengthen the effectiveness of ADB support.

I thank the Government of Mongolia for the excellent ongoing partnership with ADB, and commend the authors for this original work.

James P. Lynch
Director General, East Asia Department
Asian Development Bank

Acknowledgments

This study was conducted by a team comprising Mark R. Bezuijen, principal environment specialist, East Asia Department (EARD), Asian Development Bank (ADB); Michael Russell and Robert J. Zomer, ADB climate change consultants; and D. Enkhtaivan, ADB protected area consultant. M. R. Bezuijen, who initiated and coordinated the study, was lead author for Chapters 1, 2, 6, and 7; co-authored (with M. Russell) the executive summary and Chapter 5; and edited the report. M. Russell led the climate modeling, prepared an early draft outline of the report, was lead author for Chapters 3 and 5, and contributed to the description of ecosystems in Chapter 4. R. J. Zomer conducted the bioclimatic modeling and was lead author for Chapter 4. D. Enkhtaivan prepared spatial data, provided local knowledge, and contributed to the study analyses and descriptions of ecosystems. The study was financed by ADB through its Climate Change Fund.

The study includes unpublished data and preliminary climate modeling prepared in 2016–2019 for two ADB-funded projects. Data and map layers for Khuvsgul Lake National Park (KLNP) on park zoning, visitor numbers, ecological values, and tourism were compiled under the Integrated Livelihoods Improvement and Sustainable Tourism in Khuvsgul Lake National Park Project (Grant 9183-MON). This data collection was led by D. Enkhtaivan, consultant for the project in 2017–2019. Climate modeling and stakeholder surveys were led by M. Russell during the preparation of Mongolia's Sustainable Tourism Development Project (Loans 3787/3788-MON). Grant 9183-MON was implemented with $3.0 million in grant funding from the Japan Fund for Poverty Reduction (JFPR). Loans 3787/3788-MON were prepared with grant funding of $1.1 million from the JFPR and $0.076 million from ADB's Climate Change Fund. M. R. Bezuijen led the design and coordination of both projects.

The following agencies provided, and gave permission to use, data for the study: the Ministry of Environment and Tourism of Mongolia (spatial data on the KLNP boundary, zones, tour camps, and herder camps); the National Agency for Meteorology, Hydrology and Environmental Monitoring of Mongolia (KLNP weather data); and The Nature Conservancy (spatial files of ecosystem mapping, cited throughout this report as Heiner et al. 2017).

The authors thank the following for their review of the report, field coordination, and/or logistical support:

- **Ministry of Environment and Tourism:** E. Sansarbayar (director general, Department of Protected Area Management), S. Bayasgalan (director general, Department of Tourism Policy Coordination), and personnel of the KLNP Administration.
- **Grant 9183-MON:** Project implementation unit staff, including T. Erdenejargal (project coordinator), D. Bayarmagnai (monitoring and evaluation officer), and B. Munguntulga (protected area specialist).
- B. Oyunmunkh (climate change consultant); and M. Nyamkhuu (researcher, Institute of General and Experimental Biology, Mongolian Academy of Sciences).

M. R. Bezuijen thanks ADB's Qingfeng Zhang (director, Environment, Natural Resources, and Agriculture Division [EAER], EARD) for support, guidance, and review of the report; Maria Pia Ancora (senior urban development specialist, Central and West Asia Department) for discussions on the study concept; and Michael Brian R. Bonilla (financial management officer, EAER), Ongonsar Purev (senior environment officer, Mongolia Resident Mission), Arghya Sinha Roy (senior climate change specialist, Sustainable Development and Climate Change Department), and Noreen Joy N. Ruanes (senior operations assistant, EAER) for their assistance.

Critical comments from two external peer reviewers (Clyde E. Goulden and one anonymous reviewer) and two ADB peer reviewers (Charles Andrew Rodgers and Francesco Ricciardi) improved the draft report. Joy Quitazol-Gonzalez facilitated the production of the final report, including final editing, proofreading, design, and publication.

Abbreviations

ADB	Asian Development Bank
AWI	aridity–wetness index
COVID-19	coronavirus disease
CSO	civil society organization
GDD	growing degree days
KLNP	Khuvsgul Lake National Park
KSSPA	Khoridol Saridag Strictly Protected Area
PET	potential evapotranspiration
RCP	representative concentration pathway
TNP	Tunkinsky National Park
TSNP	Tengis–Shishged National Park
UTSPA	Ulaan Taiga Strictly Protected Area

Weights and Measures

°C	degrees Celsius
ha	hectare
km	kilometer
km^2	square kilometer
m	meter
mm	millimeter

Executive Summary

Protected areas, such as national parks and reserves, form the basis of most national and global efforts to conserve biodiversity. They help to maintain natural living systems and the ecological processes that life depends on. Protected areas provide many benefits to human society. In the context of sustainable development, the management of protected areas is an important component of nature-based approaches to protect natural resources and "one health" for nature and people, including resilience to wildlife-related diseases.

Climate change threatens to undermine the benefits of protected areas. Mongolia supports a large network of protected areas with global biodiversity values, but is experiencing some of the fastest rates of climate change in the world. National plans are underway to expand the protected area network and promote nature-based tourism to improve the livelihoods of impoverished rural communities; yet, little information is available to support policy makers and protected area managers on how to integrate climate change within protected area planning.

This study presents the first quantitative assessment of the potential impacts of climate change on a protected area in Mongolia. It examines the projected impacts of climate change on three dimensions—biodiversity, livelihoods, and tourism—for one of Mongolia's largest and most visited protected areas, Khuvsgul Lake National Park (KLNP).

The KLNP is located in northern Mongolia and covers 11,800 square kilometers (km²). It encompasses Khuvsgul Lake, the largest source of fresh water in the country. The climate is characterized by cold, dry winters; mild, windy summers; high temperature fluctuations; and low precipitation. Landforms comprise mountains and valleys around Khuvsgul Lake (located at 1,645 meters elevation), with a large elevational range in the park (extending over 1,800 meters) that contributes to a variety of forest, steppe (grassland), and wetland habitats. Khuvsgul Lake has retained its near-pristine water quality and, like other high-altitude lakes in cold climates, has naturally low nutrient levels and is extremely sensitive to pollution. It is the only lake in the world surrounded by permafrost, a frozen sublayer of soil that is vital in maintaining soil moisture and vegetation growth. The park supports relatively small human populations, and the dominant livelihood is herding. Major threats to the park are excessive livestock grazing and unmanaged tourism, which have damaged large areas of vegetation and soil and are polluting Khuvsgul Lake.

The study approach comprised climate modeling, supplemented by stakeholder consultations. Local meteorological data were compiled to analyze weather trends over time, and publicly available climate models were applied to develop climate projections to the year 2050. A global model (Metzger et al. 2013) was applied to categorize KLNP into "bioclimatic zones," in which each zone represents a unique combination of climatic and environmental conditions that different plant species live within. The zones are divided into finer strata (layers), representing more detailed combinations of temperature, precipitation, and other parameters essential for plant growth. The climate models applied for the study are based on data from 1960 to 1990, and this time span represented the "baseline climate" from which to compare projected changes to 2050. To signify biodiversity values, a surrogate indicator, "ecosystems," was used. Ecosystems are distinct landscape units, which represent different plant and animal communities and their unique living conditions. They are an appropriate unit for this study to assess change over time, given the large size of KLNP. Ecosystems in KLNP were mapped from an existing ecosystem classification (Heiner et al. 2017).

The climate of KLNP has changed significantly over the past 50–60 years. Between 1963 and 2016, mean, minimum, and maximum temperatures in spring and summer increased by over 0.3°C/decade, and the maximum annual temperature of Khuvsgul Lake increased from 14°C to 18°C. There was no measurable change in total annual precipitation during this period, but since 1980, the number of storm events has almost doubled. Residents, tour operators, and staff of government agencies have reported changes in weather patterns consistent with these trends, including longer and hotter summers, fewer light rains (which are beneficial as they soak the earth), and more intense storms and flooding. Other studies have confirmed that the permafrost layer is melting because of rising temperatures and soil damage from human activity.

Substantial further changes in the climate are expected. By 2050, mean annual air temperatures of KLNP are estimated to have risen by 2.4°C–2.9°C, compared with the 1960–1990 baseline. When high-risk climate models are applied, increases of 5.0°C or more are foreseen for some regions of KLNP. Small increases in summer and winter precipitation are projected, and rates of evaporation will increase. Overall, the climate is becoming warmer and drier, but is likely to differ between areas because of the park's diverse topography. There will be more unpredictability and variability in weather within and between seasons. Changes in temperature are projected to be smallest at lower elevations in the park (in the floodplains and hills east of Khuvsgul Lake) and highest in the mountains east and west of Khuvsgul Lake. Increases in precipitation will be most pronounced in the eastern mountainous areas of KLNP.

Under the baseline climate period (1960–1990), the KLNP encompassed nine strata in three global bioclimatic zones: (i) extremely cold and wet zone (about 3% of KLNP), (ii) extremely cold and mesic (dry) zone (58%), and (iii) cold and mesic zone (39%). These categories reflect the generally cold and dry conditions in the park. The KLNP supports 15 types of ecosystems, comprising high-elevation alpine habitats, forest and steppe ecosystems at lower elevations, and Khuvsgul Lake. The study found a close correlation between the distribution of bioclimatic zones and ecosystems in KLNP. This confirms that vegetation communities in KLNP are associated with specific climatic and environmental conditions.

By 2050, the KLNP is projected to have undergone a profound change in bioclimatic conditions due to climate change. The mean elevation of all bioclimatic zones is projected to shift markedly upward. The extremely cold and wet zone, which is restricted to the highest elevations, and some strata of the extremely cold and mesic zone are projected to disappear from the park. The area encompassed by the cold and mesic zone—relatively the warmest zone of the park—will have almost doubled and will encompass over 82% of KLNP. New bioclimatic strata that currently do not occur in KLNP will have entered the park and displaced existing strata. Overall, about 10,983 km² (93%) of KLNP will have shifted to an entirely different set of bioclimatic conditions not previously experienced in that location.

For biodiversity, these changes will almost certainly cause severe and irreversible impacts on the composition of plant and animal communities, individual species, and the conservation values of KLNP. Ecosystems adapted to warmer conditions and that presently only occur at lower elevations in KLNP will expand and displace the cold-adapted ecosystems of higher elevations. The treeless areas of the alpine ecosystems are likely to become populated by trees, displacing alpine plants. The area of three categories of alpine ecosystems in KLNP is projected to decline by 87%–92%. At least one rare, high-altitude plant species may become locally extinct. At lower elevations, closed forest ecosystems will be replaced by drier, open forests and steppes, which will increase the exposure of soil and permafrost to further damage and drying. Ecosystems of riverine forests and meadows may be replaced by shrub and steppe ecosystems. The KLNP supports populations of large grazing mammals, and the decline of closed forests will reduce the thick cover and rich feeding resources they rely on.

For Khuvsgul Lake, warmer waters and the increased frequency and intensity of storm events will compound the impacts of livestock waste and tourism on water quality, by promoting favorable conditions for algal blooms and increased transfer of nutrients to the lake from runoff. Changes in the seasonal temperature regime of the lake and the seasonal volume of water input from streams and rainfall, combined with water pollution, are likely to impact the communities of aquatic invertebrates and fish, which depend on high water quality and are adapted to the lake's hydrology. The impacts of high nutrient levels and climate on lake water quality and ecology will be most severe in small, semi-enclosed bays (which are numerous around the lake), especially those occupied by tour camps. Many species are unique to Khuvsgul Lake and some, including the Khuvsgul grayling (a fish species found nowhere else), are already threatened by overfishing or water pollution. There are no nearby lakes for these communities to shift to, and species unable to adapt to the new climatic conditions may become extinct.

For herding livelihoods, all areas of the park used for livestock grazing (which are mainly around Khuvsgul Lake) are projected to undergo changes in bioclimatic strata that will result in warmer and drier conditions and a shift from forested to more open ecosystems. High livestock grazing pressures have already resulted in damage to pasture and stream banks, declining soil fertility, increased permafrost melt, and fire risk. Climate change will compound these impacts, in a cycle leading to greater environmental damage. Herding households are the most impoverished residents of KLNP, and they already have a low inherent resilience to climate change because of limited opportunities for income diversification. Declining pasture productivity under climate change will increase the vulnerability of herders and livestock through reduced availability of livestock winter fodder, weaker condition of livestock, and reduced opportunity to earn income from the sale of meat and dairy products. To supplement livestock resources, hunting and fishing may increase, placing further pressures on biodiversity.

For tourism, climate change may result in damage to park infrastructure and higher costs of operation and maintenance for the government and tour operators, risks to visitor safety, and reduced visitor experiences. The melting of permafrost has already resulted in localized land subsidence and damage to some buildings in KLNP. More frequent or intense weather events (e.g., storms, fire, rapid changes in temperature) will increase the risk of damage to roads and tour camps from flooding, wave action on Khuvsgul Lake, rapid freeze–thaw cycles, and hazards to residents and visitors. For Khuvsgul Lake, warming lake temperatures, shorter winters, and continued pollutant inputs from tourism and livestock are likely to degrade the lake conditions. Warming conditions, combined with low existing sanitation standards, may increase the transmission of infectious diseases. Khuvsgul Lake is the centerpiece of tourism in KLNP, and such impacts are likely to result in visitor complaints. Declines in tourism due to these various issues would reduce the opportunity for residents to benefit from tourism. This is significant, as tourism—if managed sustainably—provides an opportunity to strengthen the resilience of communities to climate change through income diversification, and is one of the few livelihood opportunities compatible with the conservation objectives of KLNP.

To build resilience to climate change in KLNP, at least three approaches are required: address existing threats to biodiversity, improve habitat connectivity, and strengthen park management. In the short term, the highest priority is to address the impacts of unsustainable livestock grazing and unmanaged tourism. Given the multiple-use objectives of KLNP, a multisector approach is critical to benefit conservation and livelihoods. Measures are recommended for improved livestock and pasture management, tourism planning, and waste management. For habitat connectivity, the KLNP is located in a landscape that is well suited to transboundary conservation. Nearby regions support a mosaic of protected areas and limited development. With effective planning, these attributes can enable some KLNP ecosystems to shift northward or to higher elevations as bioclimatic conditions in KLNP become unsuitable. As for park management, local agencies are already under-resourced and there is presently no planning for climate change. Institutional reform, revision of the KLNP management plan, new long-term management targets, and increased capacity and resources are required to effectively address existing issues and plan for climate change.

The Asian Development Bank has provided support for livelihoods, sustainable tourism, and conservation in KLNP through two projects being implemented between 2016 and 2024. The projects build on a large platform of national and international support for KLNP that has been provided by other agencies for scientific research, livelihoods, and park management. These various efforts have helped address some of the measures required to build resilience to climate change in KLNP, but the needed actions are beyond the scope of any single project or agency. Collaboration and coordination between national and international agencies, focused especially on the park's ecological values most threatened by climate change, as well as on herding livelihoods and tourism, will be critical to achieving the effective management of KLNP under climate change.

Executive Summary (Mongolian)

Үндэсний болон олон улсын түвшинд биологийн олон янз байдлыг хамгаалах хүчин чармайлтын бааз суурийг гол төлөв байгалийн цогцолборт газар, байгалийн нөөц газар зэрэг тусгай хамгаалалттай газар нутгууд бүрдүүлэхийн зэрэгцээ амьдралыг тэтгэгч амьд байгалийн тогтолцоо, экологийн процессийг хадгалахад тусалдаг. Тусгай хамгаалалттай газар нутгаас хүний нийгмийн хүртдэг өгөөж арвин билээ. Тогтвортой хөгжлийн агуулгын хүрээнд байгалийн нөөц, байгаль болон хүний "нэгдмэл эрүүл мэнд"-ийг хамгаалах, үүний дотор зэрлэг ан амьтадтай холбоотой өвчлөлд тэсвэртэй байх чадварыг бий болоход чиглэсэн байгаль экологийн аргачлалын чухал бүрэлдэхүүн хэсэг бол тусгай хамгаалалттай газар нутгийн менежмент юм.

Уур амьсгалын өөрчлөлт тусгай хамгаалалттай газар нутгийн үр өгөөжид аюул занал учруулж байна. Дэлхийд үнэ цэнэтэйд тооцогддог биологийн олон янз байдлыг агуулсан тусгай хамгаалалттай газар нутгийн томоохон сүлжээ бүхий Монгол Улс уур амьсгалын өөрчлөлтөд хамгийн хурдацтай өртөж буй дэлхийн улс орнуудын нэг юм. Хөдөө орон нутгийн ядуу иргэдийн амьжиргааг дээшлүүлэх зорилгоор тусгай хамгаалалттай газар нутгийн сүлжээг өргөжүүлэх, байгалийн аялал жуулчлалыг хөгжүүлэх чиглэлээр үндэсний хөтөлбөрүүд хэрэгжиж байгаа хэдий ч тусгай хамгаалалттай газар нутгийн төлөвлөлтөд уур амьсгалын өөрчлөлтийг хэрхэн тусгах талаарх мэдээлэл бодлого боловсруулагчид болон тусгай хамгаалалттай газар нутгийн хамгаалалтын захиргааны удирдлагуудын хувьд хомс байна.

Монгол Улсын тусгай хамгаалалттай газар нутагт уур амьсгалын өөрчлөлтөөс үзүүлж болзошгүй нөлөөллийн анхны тоон үнэлгээг энэхүү судалгаагаар танилцуулж байна. Энэхүү судалгаанд Монгол Улсын хамгийн өргөн уудам бөгөөд хамгийн олон хүн зорин очдог тусгай хамгаалалттай газар нутгууудын нэг болох Хөвсгөл нуурын байгалийн цогцолборт газар (ХНБЦГ)-т уур амьсгалын өөрчлөлтөөс үзүүлэх нөлөөллийн хэтийн төлөвийг биологийн олон янз байдал, амьжиргаа, аялал жуулчлал гэсэн гурван хэмжигдэхүүний хүрээнд судлав.

ХНБЦГ нь Монгол Улсын нутаг дэвсгэрийн хойд чанад дахь 11,800 км^2 нутаг дэвсгэрт Монголын хамгийн том цэнгэг усны эх үүсвэр болох Хөвсгөл нуурыг хамран оршино. Цаг агаарын хувьд температурын хэлбэлзэл өндөртэй хүйтэн, хуурай өвөл, салхирхаг, хур тунадас багатай, аядуу зөөлөн зунтай. Газрын тогтоц нь ой, хээр (бэлчээрийн) болон намгархаг газрын амьдрах орчныг цогцлоосон Хөвсгөл нуур (далайн түвшнээс дээш 1,645 метрийн өндөрт оршдог)-ыг тойрсон уулс, хөндий, байгалийн цогцолборт газар нутаг дэвсгэр дэх өндөрлөг нуруудын (1,800 м-ээс дээш өргөгдсөн) хэв шинжийг агуулна. Хөвсгөл нуурын ус онгон цэнгэг чанараа хадгалж үлдсэн бөгөөд хүйтэн уур амьсгалтай өндөрлөг бүсэд орших бусад нууруудын нэгэн адил байгалийн тэжээллэг чанар багатай, бохирдолд маш эмзэг мэдрэмтгий юм. Тэрээр хөрсний чийгшил, ургамлын ургалтыг хадгалахад чухал үүрэг гүйцэтгэдэг мөнх цэвдэг хөрсөөр хүрээлэгдсэн дэлхийн цорын ганц нуур юм. Амьжиргаандаа мал аж ахуй зонхилон эрхэлдэг харьцангуй цөөн тооны хүн ам тус байгалийн цогцолборт газарт амьдардаг. Малын бэлчээрийн даац хэтрэлт, аялал жуулчлалын замбараагүй үйл ажиллагаа нь тус байгалийн цогцолборт газрын хүрээлэн буй орчинд томоохон аюул занал учруулж, газар нутгийн ихээхэн хэсгийн ургамал, хөрсийг сүйтгэн, Хөвсгөл нуурыг бохирдуулж байна.

Уур амьсгалын загварчлалыг оролцогч талуудтай зөвлөлдөх аргачлалтай хослуулан судалгааны арга зүйд ашиглав. Хугацаанаас хамаарсан цаг уурын чиг хандлагыг шинжих зорилгоор орон нутгийн цаг агаарын мэдээг нэгтгэн боловсруулж, уур амьсгалын төлөв байдлыг судлах нийтлэг загварчлалуудыг ашиглан 2050 он хүртэлх цаг уурын хэтийн төлөвийг боловсруулаа. Олон улсын загварчлал (Metzger et al. 2013)-ын дагуу бүс бүрт нь өөр өөр төрөл зүйлийн ургамал амьдардаг уур амьсгал, байгаль орчны өвөрмөц хослолыг цогцлоосон "био-уур амьсгалын бүс" гэдэг ангилалд ХНБЦГ-ыг хамруулсан. Уг бүсүүд нь температур, хур тунадас болон ургамлын өсөлтөд чухал шаардлагатай бусад үзүүлэлтүүдийн нарийвчилсан хослолыг хамруулсан нарийн давхаргууд (үеүд)-ад хуваагдана. Энэхүү судалгаанд ашигласан уур амьсгалын загварчлалд 1960–1990 оны мэдээллийг "суурь үзүүлэлт" болгон авсан бөгөөд түүнийг 2050 он хүртэлх уур амьсгалын төлөвийн өөрчлөлттэй харьцуулан судлав. Биологийн олон янз байдлыг томъёолоходоо орлуулах үзүүлэлт (surrogate indicator) болон "экосистем"-ийн ойлголтыг ашигласан болно. Экосистем нь ургамал, амьтны өөр өөр популяцууд, тэдгээрийн амьдралын өвөрмөц нөхцөлийг илэрхийлсэн ландшафтын онцгой нэгж юм. Тэрээр өргөн уудам нутгийг хамран орших ХНБЦГ-ын хувьд цаг хугацааны явцад гарсан өөрчлөлтийг үнэлэх судалгааны тохиромжтой нэгж болно. ХНБЦГ-ын экосистемийн зураглалыг одоогийн экосистемийн ангилал дээр тулгуурлан гүйцэтгэлээ (Heiner et al. 2017).

Сүүлийн 50–60 жилийн хугацаанд ХНБЦГ-ын уур амьсгал эрс өөрчлөгдсөн. 1963–2016 оны хооронд арван жил тутам хавар, зуны дундаж, хамгийн бага, хамгийн их температур 0.3°С-аар нэмэгдэж, 14°С байсан Хөвсгөл нуурын жилийн хамгийн их температур 18°С болж нэмэгдсэн байна. Дээрх хугацаанд жилийн нийт хур тунадасны хэмжээнд мэдэгдэхүйц өөрчлөлт гараагүй хэдий ч 1980 оноос хойш хүчтэй аадар борооны тоо бараг хоёр дахин олширчээ. Цаг агаарын төлөв байдал ийнхүү өөрчлөгдөж, тухайлбал зуны улирлын үргэлжлэх хугацаа урт, зундаа халуун, шиврээ бороо (газрын хөрсөнд сайн шингэдгээрээ ач холбогдолтой) ховор, хүчтэй аадар бороо, үерийн давтамж нэмэгдэж байгаа талаар оршин суугчид, тур операторууд, засгийн газрын агентлагууд мэдээлж байна. Хүний үйл ажиллагаанаас үүдэлтэй температурын өсөлт, хөрсний эвдрэлийн улмаас мөнх цэвдэгт давхарга хайлж байгааг мөн өөр нэгэн судалгаагаар тогтоосон байдаг.

Цаашид уур амьсгалын томоохон өөрчлөлтүүд гарах хандлагатай байна. 2050 он гэхэд ХНБЦГ-ын агаарын жилийн дундаж температур 1960–1990 оны суурь үзүүлэлттэй харьцуулахад 2.4–2.9°С-аар нэмэгдэх төлөв ажиглагдана. Өндөр эрсдэлүүдийг тооцсон уур амьсгалын загварчлалуудыг ашиглан тооцож үзэхэд ХНБЦГ-ын зарим бүс нутгуудад агаарын жилийн дундаж температур 5°С буюу түүнээс дээш хэмжээгээр нэмэгдэх төлөвтэй байна. Зуны болон өвлийн улиралд хур тунадасны хэмжээ бага зэрэг нэмэгдэх бөгөөд уурших хэмжээ мөн нэмэгдэнэ. Уур амьсгал ерөнхийдөө улам дулаарч, хуурайших байгаа боловч ХНБЦГ-ын байр зүйн тогтоцын онцлогоос хамааран нутаг дэвсгэрийн хэмжээнд харилцан адилгүй байх магадлалтай. Тухайн улиралд болон улирал хооронд цаг агаарын байдал урьдчилан таамаглахад илүү төвөгтэй, хувьсамтгай байх болно. ХНБЦГ-ын нам дор газар (Хөвсгөл нуурын зүүн талын татам болон толгод)-т температурын өөрчлөлт хамгийн бага, харин Хөвсгөл нуурын зүүн ба баруун талын ууллархаг хэсэгт хамгийн өндөр байх төлөвтэй байна. ХНБЦГ-ын зүүн талын ууллархаг нутгаар хур тунадасны хэмжээ хамгийн их нэмэгдэнэ.

Уур амьсгалын суурь үзүүлэлтийн хугацаанд (1960–1990) ХНБЦГ нь дэлхийн био-уур амьсгалын "хэт хүйтэн, чийглэг бүс" (ХНБЦГ-ын 3% -ийг хамарсан), "хэт хүйтэн, гандуу [хуурай] бүс" (58%), "хүйтэн, гандуу бүс" (39%) гэсэн гурван бүсийн есөн давхаргад хамаарч байв. Эдгээр ангиллууд нь ХНБЦГ-ын хүйтэн, хуурай нөхцөлийг ерөнхийд нь илтгэнэ. ХНБЦГ-т 15 төрлийн экосистем бүрэлдэн тогтсон бөгөөд үүнд өндөр уулын бүс, нам дор газрын ой, хээрийн экосистем, Хөвсгөл нуур багтдаг. ХНБЦГ дахь био-уур амьсгалын бүсүүд ба экосистемүүдийн тархалт хоорондоо нягт уялдаа холбоотой болохыг судалгааны үр дүнд тогтоов. ХНБЦГ-ын ургамлын популяцууд нь уур амьсгал, хүрээлэн буй орчны тодорхой нөхцлүүдтэй холбоотой болох нь үүгээр нотлогдож байна.

Уур амьсгалын өөрчлөлтийн нөлөөгөөр 2050 он гэхэд ХНБЦГ-ын био-уур амьсгалын нөхцөлд гүнзгий өөрчлөлт орох төлөв ажиглагдаж байна. Био-уур амьсгалын бүх бүсүүдийн дундаж өндөржилт огцом дээш шилжих төлөвтэй байна. Хамгийн өндөрт орших тагийн бүслүүрт орших хэт хүйтэн, чийглэг бүс, түүнчлэн хэт хүйтэн, гандуу бүсийн зарим давхарга ХНБЦГ-аас алга болно гэсэн төсөөлөл байна. Тус байгалийн цогцолборт газрын харьцангуй дулаан бүс болох хүйтэн, гандуу бүсэд хамаарах газар нутгийн хэмжээ бараг хоёр дахин нэмэгдэж, ХНБЦГ-ийн 82 гаруй хувийг эзлэх болно. Өнөөг хүртэл

ХНБЦГ-т байгаагүй био-уур амьсгалын шинэ давхарга тус байгалийн цогцолборт газарт түрэн орж, одоо байгаа давхаргыг шахан эзэлнэ. Нийтдээ ХНБЦГ-ын 10,983 км² (93%) нутаг дэвсгэр тухайн байршилд урьд өмнө байгаагүй био-уур амьсгалын эрс өөр нөхцөлд шилжсэн байна.

Биологийн олон янз байдлын хувьд эдгээр өөрчлөлтүүд нь ургамал, амьтдын популяци, бие даасан төрөл зүйлс, ХНБЦГ-ын хамгааллын үнэт зүйлсэд ноцтой бөгөөд эргэлт буцалтгүй нөлөөлөл үзүүлэх нь дамжиггүй. Одоогоор зөвхөн ХНБЦГ-ын нам дор газарт тархсан, дулаан нөхцөлд дасан зохицсон экосистем өргөжин тэлж улмаар өндөрлөг газрын хүйтэнд дасан зохицсон экосистемийг түрэн эзлэх болно. Өндөр уулын экосистемийн модгүй нүцгэн талбай ойгоор бүрхэгдэж, өндөр уулын ургамлыг орлох магадлалтай. ХНБЦГ дахь өндөр уулын экосистемийн гурван ангилалд хамаарах газар нутгийн хэмжээ 87%–92%-иар буурах төлөвтэй байна. Өндөр уулын бүсийн нэг буюу түүнээс олон төрлийн ховор ургамлын үүлдэр тухайн газар нутагт устаж үгүй болох магадлалтай. Нам дор газарт титэм нийлсэн ойн экосистем нь хуурайдуу тармаг ой, хээрээр солигдсоноор хөрс, мөнх цэвдэгт нөлөөлж, яваандаа экосистем улам бүр доройтож, хуурайших болно. Гол мөрний эргийн зурвас ой, нугын экосистем нь бут сөөг, хээрийн экосистемээр солигдох магадлалтай. ХНБЦГ-т бэлчээрийн том хөхтөн амьтдын популяци нутагшдаг бөгөөд титэм нийлсэн ойн талбай хумигдсанаар тэдний нөмөр нөөлөг, хоол тэжээлийн нөөц багасах болно.

Малын гаралтай хаягдал, аялал жуулчлалын үйл ажиллагаанаас үүдэлтэй бохирдлын зэрэгцээ хүчтэй аадар борооны эрчим, давтамж нэмэгдэх, нуурын ус бүлээсэх зэрэг хавсарсан нөлөөллийн улмаас Хөвсгөл нуурын усанд замаг үржих, гэрийн усны урсацтай хамт шим тэжээл зөөвөрлөгдөн нуурын усанд орох таатай нөхцөлийг бүрдүүлнэ. Нуурын улирлын температурын горим болон гол горхи, хур тунадсаар тэжээгдэн нуурт цутгах улирлын усны хэмжээнд орох өөрчлөлт, усны бохирдол зэрэг нь усны цэнгэг байдлаас хамааралтай бөгөөд нуурын гидрологийн нөхцөлд дасан зохицсон усны сээр нуруугүй амьтад, загасны популяцид нөлөөлөх нь дамжиггүй. Нуурын усны шим тэжээлийн түвшин болон уур амьсгалын нөхцөлөөс нуурын усны чанар, экологид үзүүлэх нөлөөлөлд нуурын эргэн тойронд олон тоогоор байдаг жижиг, хагас булан тохойнууд, ялангуяа аялал жуулчлалын баазууд байрладаг ийм булангууд хамгийн хүчтэй өртөх болно. Хөвсгөл нуурт л зөвхөн амьдардаг олон төрөл зүйл, үүний дотор Хөвсгөлийн хадран (өөр хаа ч байдаггүй загасны төрөл зүйл) хэт их загасчлал болон усны бохирдолд нэрвэгдэх аюулд учраад байна. Эдгээр популяцуудын хувьд шилжин нутагших боломжтой нуур цөөрөм ойр орчинд нь байхгүй тул уур амьсгалын шинэ нөхцөлд дасан зохицох чадваргүй төрөл зүйлүүд бүрмөсөн устаж болзошгүй.

Тус байгалийн цогцолборт газрын хэмжээнд мал аж ахуйн бэлчээрийн зориулалтаар ашиглагдаж буй нутаг дэвсгэр (ихэвчлэн Хөвсгөл нуурын эрэг орчим) бүхэлдээ био-уур амьсгалын давхаргын өөрчлөлтөд орж, илүү дулаан, хуурай орчин нөхцөл бүрдсэнээр ой мод бүхий экосистемээс илүү тармаг задгай экосистемд шилжих төлөвтэй байна. Малын бэлчээрийн ачаалал нэмэгдсэний улмаас бэлчээр, гол горхины гольдрол эвдрэн сүйрч, хөрсний үржил шим буурч, мөнх цэвдэг хайлж, гал түймрийн аюулд өртөж байна. Уур амьсгалын өөрчлөлт нь эдгээр нөлөөллийг улам хүндрүүлж, хүрээлэн буй орчинд илүү их хохирол учруулах мөчлөг рүү түлхэнэ. ХНБЦГ-ын хэмжээнд хамгийн их ядууралд өртсөн хүн ам болох малчин өрхүүдэд орлогоо төрөлжүүлэх боломж хязгаарлагдмал тул уур тэд амьсгалын өөрчлөлтөд дасан зохицох, даван туулах чадавхи султай байдаг. Уур амьсгалын өөрчлөлтийн нөлөөгөөр бэлчээрийн үржил шим буурч байгаагийн улмаас малын өвлийн тэжээлийн нөөц хомсдож, малын тарга тэвээрэг буурах, мах, сүүн бүтээгдэхүүний борлуулалтаас орлого олох боломж хумигдаж улмаар малчид болон мал сүргийн эмзэг байдал нэмэгдэхэд хүргэнэ. Мал сүргийн нөөцийг өсгөхийн тулд ан агнах, загасчлах хандлага нэмэгдэж, биологийн олон янз байдалд дарамт учруулж болзошгүй юм.

Аялал жуулчлалын хувьд уур амьсгалын өөрчлөлт нь байгалийн цогцолборт газрын дэд бүтцэд хохирол учруулах, ашиглалт, засгийн газар болон тур оператор компаниудын түвшинд засвар үйлчилгээний зардлыг өсгөх, аялагч зорчигчдын аюулгүй байдалд эрсдэл учруулж, жуулчдын сэтгэгдэлд сөргөөр нөлөөлж болзошгүй. Мөнх цэвдэг хайлсны улмаас ХНБЦГ-т газар цөмрөх, зарим барилга байгууламж эвдрэлд орох тохиолдол гарсан. Цаг агаарын үзэгдлүүд (тухайбал, хүчтэй аадар бороо, гал түймэр, температурын огцом өөрчлөлт)-ийн давтамж, эрчим нэмэгдсэний улмаас авто зам, аялал жуулчлалын баазууд үерт өртөн эвдэрч гэмтэх, Хөвсгөл нуурын давалгаа нэмэгдэх, хөлдөлт, гэсэлтийн мөчлөгийн хурд нэмэгдэх, оршин суугчид болон аялагч жуулчид аюулд өртөх эрсдэлийг нэмэгдүүлж байна. Хөвсгөл нуурын хувьд нуурын температур нэмэгдэх, өвлийн улирлын үргэлжлэх хугацаа богиносох, аялал жуулчлал, мал аж ахуйн үйл ажиллагаанаас үүдэлтэй бохирдол нэмэгдэх зэрэг нь нуурын нөхцөл байдлыг доройтуулж болзошгүй юм. ХНБЦГ-ын аялал жуулчлалын гол цөм нь Хөвсгөл нуур тул ийм нөлөөлөл нь гомдол гарах шалтгаан болно. Дулаарлын нөхцөл байдал, одоо байгаа ариун цэврийн байгууламж стандартын шаардлага хангаагүй байгаа зэрэг нь халдварт өвчин тархах эрсдэлийг нэмэгдүүлэх магадлалтай. Энэ мэт олон талын асуудлаас үүдэн аялал жуулчлал буурвал нутгийн иргэдийн аялал жуулчлалаас хүртэх буруулна. Аялал жуулчалалын тогтвортой менежментийг хэрэгжүүлснээр иргэдийн орлогыг төрөлжүүлэх, улмаар тэдний уур амьсгалын өөрчлөлтөд тэсвэртэй байдлыг бэхжүүлэх боломжийг бүрдүүлэх бөгөөд энэ нь ХНБЦГ-ын байгаль хамгаалах зорилттой нийцсэн амьжиргааг дэмжих цөөн боломжуудын нэг юм.

ХНБЦГ-т уур амьсгалын өөрчлөлтөд тэсвэртэй байдлыг бий болгохын тулд биологийн олон янз байдалд учирч буй аюул заналыг арилгах, амьдрах орчны харилцан уялдаа холбоог сайжруулах, байгалийн цогцолборт газрын менежментийг бэхжүүлэх зэрэг чиглэлээр үйл ажиллагаагаа явуулах шаардлагатай. Малын бэлчээрийн даац хэтрүүлэлт, аялал жуулчлалын замбараагүй, хяналтгүй үйл ажиллагаанаас учирч буй нөлөөллийг арилгах асуудал богино хугацаанд нэн тэргүүн ээлжийн зорилт болон тулгарч байна. ХНБЦГ-ыг олон талаар ашиглах зорилтын хүрээнд байгаль хамгаалал, амьжиргаанд өгөх үр өгөөжийг нь хүртэхийн тулд салбар дундын аргачлал нэн чухал юм. Мал аж ахуй, бэлчээрийн менежмент, аялал жуулчлалын төлөвлөлт, хог хаягдлын менежментийг сайжруулахад чиглэсэн арга хэмжээ авч хэрэгжүүлэхийг зөвлөж байна. Амьдрах орчны уялдаа холбооны хувьд ХНБЦГ нь хил дамнасан байгаль хамгаалах үйл ажиллагаанд тохирсон ландшафтад байрладаг. Зэргэлдээ орших бүс нутгууд нь тусгай хамгаалалттай газар нутгийн хосолмол хэв шинж болон хязгаарлагдмал хөгжлийг дэмжих бөгөөд үр дүнтэй төлөвлөлтийг хийснээр эдгээр хэв шинжүүд нь ХНБЦГ-ын био-уур амьсгалын нөхцөл тохиромжгүй болох нөхцөлд түүний зарим экосистем умард зүгт эсвэл өндөрлөг бүс рүү шилжих боломжийг бүрдүүлнэ. Байгалийн цогцолборт газрын менежментийн хувьд орон нутгийн агентлагуудын хүч нөөц дутмаг, уур амьсгалын өөрчлөлтийн төлөвлөлт одоогоор хийгдээгүй байна. Тулгамдаж буй асуудлуудыг үр дүнтэй байдлаар шийдвэрлэх, уур амьсгалын өөрчлөлтийг төлөвлөхийн тулд институцийн шинэчгэл хийх, ХНБЦГ-ын менежментийн төлөвлөгөөг шинэчлэх, урт хугацааны шинэ зорилтуудыг тодорхойлох, хүчин чадал, нөөцийг нэмэгдүүлэх шаардлагатай байна.

ХНБЦГ-т амьжиргааг дэмжих, тогтвортой аялал жуулчлалыг хөгжүүлэх, байгаль хамгаалах чиглэлээр дэмжлэг үзүүлэх хоёр төслийг Азийн хөгжлийн банк 2016–2024 онуудад хэрэгжүүлж байна. ХНБЦГ-т эрдэм шинжилгээ судалгаа, амьжиргааг дэмжих, байгалийн цогцолборт газрын менежментийн чиглэлээр бусад байгууллагуудын дэмжлэгтэйгээр үндэсний болон олон улсын түвшинд бий болгосон дорвитой томоохон платформ дээр суурилан эдгээр төслүүдийг хэрэгжүүлж байгаа билээ. Дээр дурдсан төрөл бүрийн арга хэмжээ нь ХНБЦГ-т уур амьсгалын өөрчлөлтөд тэсвэртэй байдлыг бий болгоход шаардлагатай зарим асуудлыг шийдвэрлэхэд түлхэц үзүүлсэн хэдий ч цаашид аливаа нэг төсөл, аль нэг агентлагийн үйл ажиллагааны хамрах хүрээнээс илүү, бусад арга хэмжээг авч хэрэгжүүлэх шаардлагатай юм. Үндэсний болон олон улсын байгууллагуудын хамтын ажиллагаа, уялдаа холбоог нэн ялангуяа уур амьсгалын өөрчлөлтөд хамгийн ихээр өртөж буй тус байгалын цогцолборт газрын экологийн үнэ цэнэ, түүнчлэн малчдын амьжиргаа, аялал жуулчлал зэрэг асуудалд чиглүүлэх явдал уур амьсгалын өөрчлөлтийн нөхцөлд ХНБЦГ-ын үр дүнтэй менежтентэд чухал үүрэг гүйцэтгэнэ.

1 Introduction

Protected areas form the basis of most national and global efforts to conserve biodiversity. Almost 15% of the earth's land surface and inland waters are protected (UNEP–WCMC, IUCN, and NGS 2018): these areas cover a vast network of habitats that helps maintain populations of plants and animals and the ecological processes that life depends on. Protected areas benefit human society through the protection of essential services such as water supply and flood regulation, and as places of recreation and well-being. Following the outbreak and global spread of the coronavirus disease (COVID-19) in 2019 and 2020, there has been renewed awareness of the need to maintain healthy ecosystems to reduce the risk of disease transmission from wildlife to people (Everard et al. 2020). Protected areas are important in this regard, as they contribute to ecosystem health, help conserve the genetic diversity that provides resilience to disease, and provide a buffer between human and wildlife populations. For these reasons, the management of protected areas is a critical component of nature-based solutions to achieve sustainable development. As human populations continue to grow, the benefits of protected areas for nature and people will become increasingly important.

Climate change caused by human development is threatening much of life. All known species live within specific ranges of temperature, humidity, and other climate parameters, and are adapted to these for their survival. Climate change is altering these conditions at rates faster than most species can adapt. This is impacting species and ecological processes. As conditions become warmer and drier, some species are shifting to higher elevations or latitudes to remain within the cooler and wetter conditions to which they are adapted. For others, which are unable to make such movements or for which there is no other habitat, populations are disappearing. These effects are causing the extinction of species, changes in the distribution of species, and impacts to entire living systems (Parmesan and Yohe 2003; Foden et al. 2013). Many species are already declining in number due to existing threats, including habitat loss, pollution, hunting, and wildlife trade, and climate change is compounding these impacts (Opdam and Wascher 2004). Warmer conditions and other changes are also increasing the risk of disease outbreaks and disease spread, presenting greater risks to people and wildlife populations.

Protected areas, if managed effectively, can play a critical role in reducing the impacts of climate change on nature and people (Watson et al. 2012). The protection of large areas of intact natural landscapes can help maintain ecological processes; provide habitats at different elevations and latitudes for plants and animals to shift to; and buffer the damage from increased storms, fires, and other extreme weather. In practice, however, achieving these benefits presents major challenges for policy makers and land managers. Most protected areas face existing pressures from human activities. Protected areas designated for multiple use, such as national parks, and/or which support human populations, require balanced planning to support conservation, livelihoods, and development; yet these priorities are often not compatible. Conservation planning requires an understanding of the living requirements of different species, but such information is often limited for most species. Management needs different skill sets, inclusive dialogue, and land use planning; however, the institutional frameworks, capacity, and/or resources are often lacking.

Climate change compounds these challenges and threatens the integrity of protected areas. Climate space is shifting, yet protected areas have fixed boundaries. As populations of plants and animals shift locations or decline in response to climate change, management priorities and actions need to be revised to reduce these impacts. To develop such actions, an understanding of the site-specific impacts of climate change is needed. There is a large global literature on the impacts of climate change on protected areas (e.g., Geyer et al. 2017; Elsen et al. 2020), but few site-specific assessments have been conducted for protected areas in Asia. Such assessments are urgently needed to help plan for climate change.

This study presents the first quantitative assessment of the projected impacts of climate change on a protected area in Mongolia. It examines potential climate impacts on three dimensions—biodiversity, livelihoods, and tourism—for one of Mongolia's largest and most visited protected areas, Khuvsgul Lake National Park (KLNP). Measures to help strengthen the park's resilience to climate change are identified.

Protected Areas and Climate Change in Mongolia

Mongolia's network of national protected areas comprises over 31.1 million hectares (ha) (about 19.8% of Mongolia's area) of land and water (MET 2019). This network encompasses mountains, plains, and desert wilderness; supports threatened species of plants and animals; and contains some of the largest remaining wildlife populations in Asia. It is of global importance for biodiversity conservation. Nomadic herding traditions, largely unchanged for thousands of years, continue across most of Mongolia, including within its protected areas.

Threats to Mongolia's biodiversity and challenges for protected area management include habitat loss and damage from livestock overgrazing, mining, logging, fire, unregulated tourism, pollution, and hunting (Chimed-Ochir et al. 2010). Many protected areas support rural communities who suffer from high levels of poverty; have few opportunities for income; and depend on herding, collection of timber and plants, and hunting for subsistence. Most protected area agencies have insufficient staff, equipment, or technical and financial resources for effective management, and national budgets for biodiversity conservation are small (Munkhchuluun and Chimeddorj 2013). To help address these issues, the government has prioritized the development of nature-based tourism to improve livelihoods and contribute to the financing of protected areas. Plans are underway to expand the national protected area network to 47 million ha (about 30% of Mongolia's total land area) by 2030 (MET 2019) and establish supporting infrastructure within and around protected areas, including roads, airports, and visitor and sanitation facilities (Government of Mongolia 2020).

Mongolia is experiencing some of the highest rates of climate change in the world (Dagvadorj, Batjargal, and Natsagdorg 2014; Dashkhuu et al. 2015). Observed changes since the 1950s include a decline in water resources and increased frequency of floods, drought, and fire (Angerer et al. 2008; Batima et al. 2005; Hessl et al. 2016). Climate change is causing changes in the distribution of vegetation communities (e.g., De Grandpré et al. 2011; Liancourt et al. 2015) and, together with livestock overgrazing, fire, drought, and pest damage, has led to a decline in vegetation cover and productivity (e.g., Chu and Guo 2012; Liu et al. 2013; Tian et al. 2014; Bayarjargal et al. 2019) and the melting of permafrost, a frozen sublayer of soil critical for maintaining soil moisture and vegetation and reducing fire risk (Sharkhuu et al. 2007).

Climate change is widely acknowledged as a threat to Mongolia's biodiversity and protected areas (Chimed-Ochir et al. 2010). Preliminary assessments of the potential impacts of climate change on biodiversity have been conducted for some river basins and regions (e.g., WWF Mongolia Programme Office 2011; Simonov, Goroshko, and Tkachuk 2018) and species (e.g., Ocock et al. 2006; Chimed-Ochir et al. 2010; Singh and Milner-Gulland 2011). The need to consider climate change in the design and planning of protected areas has been emphasized (Government of Mongolia 2015; Heiner et al. 2019), yet limited guidance is available for policy makers and land managers on how to plan for climate change. There is a need for studies that assess the potential impacts of climate change on protected areas and identify measures, tailored to local conditions, that can be integrated into protected area management plans (Chimed-Ochir et al. 2010).

The KLNP is ranked as one of Mongolia's most important national parks for conservation (Batsukh and Belokurov 2005). It supports Mongolia's largest freshwater resource, Khuvsgul Lake, and is a national priority for tourism development (Chapter 2). Climate change is already impacting KLNP, yet the extent of the impacts and risks is unclear—this hinders the opportunity to help build resilience to climate change (Goulden and McIntosh 2018).

Objectives of the Study

The study aimed to (i) identify, through modeling, the extent of potential climate change in KLNP; (ii) conduct a preliminary assessment of the potential impacts of climate change on three dimensions—biodiversity, livelihoods, and tourism; and (iii) identify measures to help build resilience to these potential impacts.

Approach

The approach comprised modeling of projected changes in climate and biodiversity values, supplemented by stakeholder consultations. A surrogate indicator, "ecosystems," was used to represent biodiversity values. Ecosystems are distinct landscape units, which represent different plant and animal communities and their unique living conditions. The application of ecosystems for the study, rather than an assessment of individual species of plants and animals, is appropriate, given the KLNP's large size (Chapter 2), the importance of identifying landscape-level trends, and the lack of sufficient ecological data for most species. Modeling at the ecosystem level helps identify broad trends under climate change—e.g., the "alpine barren" ecosystem represents high-elevation flora and fauna and their associated habitats.

Meteorological data from KLNP, available for the period 1963–2016, were compiled and analyzed for trends. Climate change projections to 2050 were developed based on these data and parameters downloaded from global climate models (Hijmans 2015; Hijmans et al. 2005). The year 2050 was applied for modeling as this is consistent with the time frame for national development planning (Government of Mongolia 2020) and provides a suitable medium-term target to plan and implement measures to build resilience to climate change.

Five parameters used in climate science were applied to measure the current climate and projected change: air temperature, precipitation, potential evapotranspiration (PET), aridity–wetness index (AWI), and growing degree days (GDD) (see Glossary). Ecosystems within and near KLNP were identified on the basis of an existing regional classification for northern Mongolia (Heiner et al. 2017). A bioclimatic map of KLNP, based on a global model (Metzger et al. 2013), was developed. The map divides KLNP into zones and strata (layers) that capture the full range of climatic and environmental conditions in the park to which vegetation communities are adapted. Different zones and strata are occupied by different vegetation communities. The climate change projections were applied to this map. A new map of the bioclimate in KLNP by 2050 was then developed, and projected changes in the distribution, elevation, and area of each ecosystem by 2050 were calculated.

The results were used to infer impacts on biodiversity, livelihoods (especially herding, the dominant livelihood in KLNP), and tourism. Measures to address these impacts and promote resilience to climate change were developed. These measures build upon management activities at KLNP by the government, the Asian Development Bank (ADB), and other agencies (Chapters 2 and 7).

Limitations of the study. The study represents a preliminary, landscape-level assessment of climate impacts on KLNP. Modeling was limited by the quality and resolution of available map data, especially for vegetation, geology, soils, and permafrost. The methodology is discussed in the Appendix.

Structure of the report. Chapter 2 gives an overview of KLNP, and Chapter 3 describes the KLNP's current and projected climate. In Chapter 4, the bioclimatic zones and ecosystems of KLNP under the current climate are identified. Chapter 5 contains an assessment of the projected impacts of climate change, based on the baseline data and projections described in Chapters 3 and 4. Chapter 6 presents measures to build resilience to climate change, and Chapter 7 summarizes support by ADB for livelihoods and tourism in KLNP.

2 Khuvsgul Lake National Park

Khuvsgul Lake National Park (KLNP) is located in Khuvsgul *aimag* (province), northern Mongolia (Figure 1), and was established in 1992. The park covers 11,800 square kilometers (km²) (1.18 million ha) and extends 160 kilometers (km) north–south and 160 km east–west (N51.3°–51.7°; E99.9°–102.2°). The northern boundary of the park is located along the international border between Mongolia and the Russian Federation. Elevations range from 1,645 meters (m) to 3,491 m. The dominant landforms in KLNP are a large lake (Khuvsgul Lake), plains, hills, and mountains. High mountain ranges are situated north and west of Khuvsgul Lake. East of Khuvsgul Lake, the center of the park, is dominated by low hills that rise to mountains along the eastern and southeastern park boundaries. Khuvsgul Lake is the largest freshwater lake in Mongolia. It is 136 km long and 20–40 km wide, has a total surface area of 2,760 km², and contains about 70% of Mongolia's fresh water (Goulden et al. 2006). Most of the lake is unpolluted and has near-pristine water quality. The watershed of Khuvsgul Lake comprises about 96 small rivers and streams, which are all within KLNP. The only outflow from the lake is a small river, the Eg, which drains southward into Mongolia's largest river (the Selenge) and eventually into Baikal Lake in the Russian Federation.

Northern Mongolia is characterized by cold, dry winters and mild, windy summers (Chapter 3). Khuvsgul Lake is frozen 5–6 months of the year, generally from mid-October to April. It is the only lake in the world surrounded by permafrost, a frozen sublayer of soil or rock that occurs at different depths and thicknesses throughout KLNP (Sharkhuu et al. 2007). The presence of this permafrost and the seasonal cycles of freezing and thawing help maintain soil moisture, vegetation, and resilience to fire.

The KLNP supports globally significant biodiversity and wilderness values. The park is located at the junction of two biologically and climatically distinct regions—the Siberian taiga forests (the coniferous forests of high northern latitudes) to the north, and the temperate grasslands of the central Asian steppe (large plains of grassland) to the south. This location, combined with the wide range of elevations in the park, has resulted in a rich diversity of lake, forest, and steppe habitats. The KLNP supports rare alpine flora, large mammals, and unique communities of aquatic flora and fauna in Khuvsgul Lake that are found nowhere else, including a fish species, the Khuvsgul grayling (*Thymallus nigrescens*) (Kozhova et al. 1989; Goulden et al. 2006). Set amidst forested and snow-capped mountains, Khuvsgul Lake is sacred to Mongolians and is known locally as the "Blue Pearl" or "Mother Ocean."

The KLNP is located in five *soums* (districts): Khankh (about 6,082 km² of KLNP, or 51.7%); Tsagaan–Üür (3,328 km², 28.3%); Chandmani–Undur (918 km², 7.8%); Renchinlkhümbe (856 km², 7.3%); and Alag–Erdene (573 km², 4.9%). Two towns are located within KLNP—Khatgal and Khankh. Both are administrative enclaves excised from the park. Khankh *soum* is located entirely within KLNP, a relatively unique situation that has implications for park management and livelihoods (Chapter 6). In 2018, the population of the five *soums* was estimated at 20,231. Of this total, about 6,124 persons (30.3%) reside within KLNP: 3,302 in Khatgal and 2,822 in Khankh (National Statistical Office of Mongolia 2019). Other settlements in KLNP comprise of small, seasonal herder camps. Poverty levels are high, especially in Khankh, because of remoteness and limited opportunities for income generation. The key livelihood is livestock herding for meat, dairy, and wool products, for subsistence use and sale.

The KLNP is a site of national priority for tourism development (Government of Mongolia 2020), and tourism is the largest commercial activity in the park. Visitor numbers have increased rapidly, from 7,716 in 2004 to 74,178 in 2019 (comprising 57,696 domestic and 16,482 international visitors), with a maximum recorded annual total number of 89,652 in 2018 (KLNP Administration, unpublished data). This rapid increase has largely been due to improved road access leading to the park. In 2020, there were about 75 tour camps in KLNP, mainly on the southwest, southeast, and northeast shores of Khuvsgul Lake (Figure 1) and about 54 tour boats in operation.

Figure 1: Map of Khuvsgul Lake National Park, Showing the Areas Described in This Report

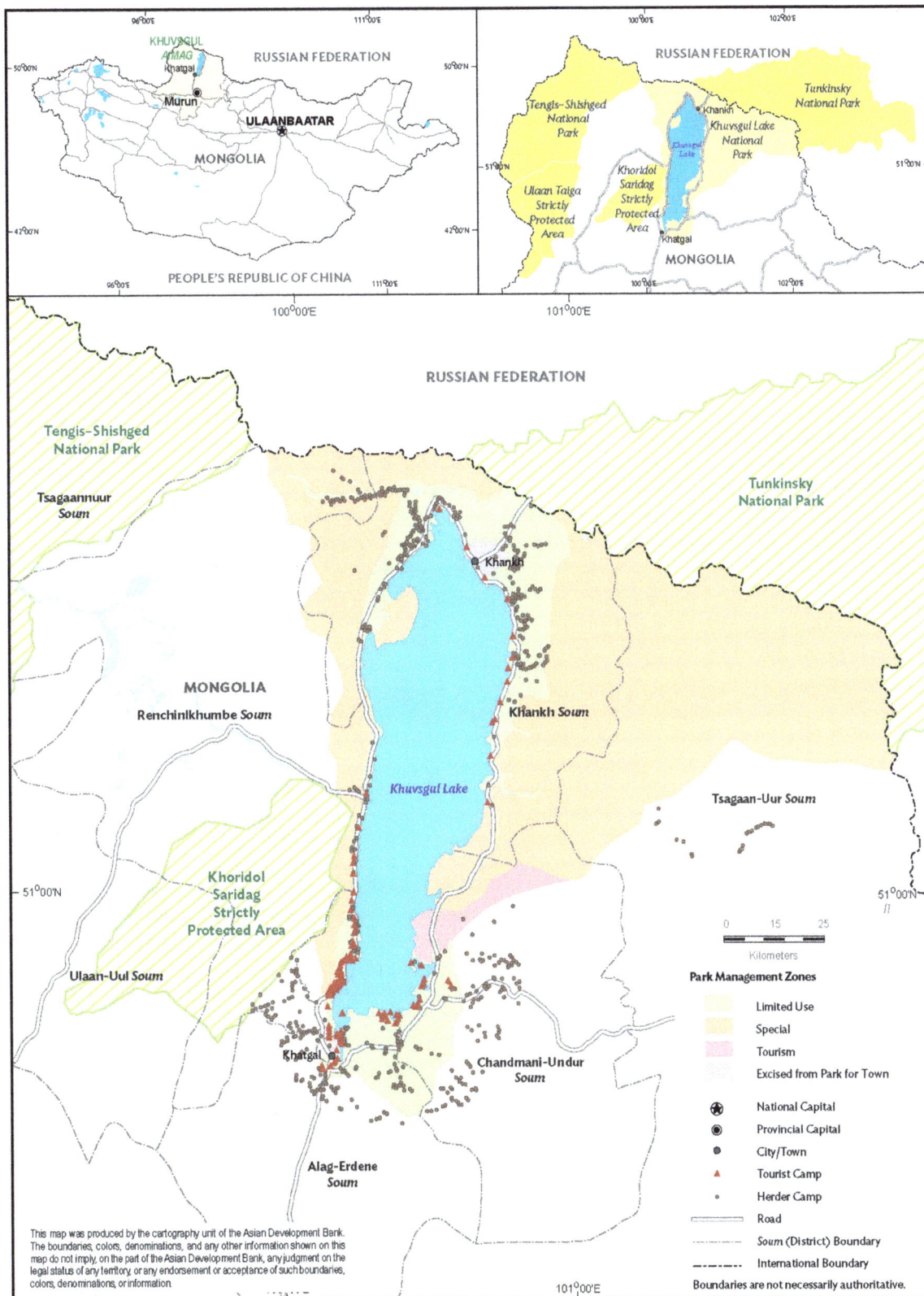

Source: Asian Development Bank.

Most tourism is from June to August (peak summer) and in March (for an ice festival; Tourism, Chapter 5). Road conditions within KLNP are poor. Khatgal and Khankh are the main entry points to the park, and most tourism is limited to the north and south shores of Khuvsgul Lake. Visitor activities include boating, camping, horseback riding, and fishing. Tourism is an important source of seasonal income for some residents.

The KLNP is managed by the KLNP Administration (located in Khatgal town) under the Ministry of Environment and Tourism. In 2020, the administration had 35 staff, comprising a director, four technical officers (for tourism and training, land management, natural resource protection and research, and legislation enforcement and monitoring), an accountant, an administration assistant, and 28 park rangers (KLNP Administration in litt.). Management of the park is also guided by the Khuvsgul Lake–Eg River Basin Administration, an agency that is also under the ministry and is part of a network of river basin authorities (Fan 2020), the Khuvsgul *aimag* government, and the five *soum* governments in the KLNP area.

The principal management document for the park is the KLNP Management Plan, 2015–2020 (MEGDT 2014b), which provides a summary of the park's conservation values, threats, and management priorities. The stated vision for the park is that "the pristine, diverse, natural, historical, cultural, and scenic resources of the lake, also imbued with spiritual values and the local nomadic traditions and folklore that support these values, are conserved through long-term preservation, fully integrated with sustainable tourism development" (MEGDT 2014b, in translation).

Under Mongolia's Law on Special Protected Areas, national parks are designated for multiple use, and regulated tourism and livelihood activities are permitted. The KLNP is divided into three management zones (Figure 1): (i) a special zone (6,820 km^2, 57.6% of the park area), where the highest level of protection is provided and human activity, except research and limited grazing of livestock, is prohibited; (ii) a tourism zone (3,122 km^2; 26.6%), including all of Khuvsgul Lake, where infrastructure (e.g., visitor centers, tour camps, car parks, roads) is permitted; and (iii) a limited use zone (1,877 km^2; 15.8%), in which herder camps, livestock husbandry, and other livelihood activities are permitted.

Other documents that contribute to the management of KLNP include an integrated water resources plan by the Khuvsgul Lake–Eg River Basin Administration (which stipulates the need to protect Khuvsgul Lake, reduce water-related health risks, and sustainably develop water resources) (MEGDT 2014a) and provincial and *soum* development plans. Management challenges for KLNP include inadequate technical and financial resources and overlapping jurisdictions between agencies (Strengthening Park Management, Chapter 6).

Considerable scientific research and support for management have been implemented at KLNP. The park has been designated as a long-term ecological research site (Goulden et al. 2000), and studies have been conducted on climate, hydrology, ecology, and livelihoods (e.g., Goulden et al. 2006; Goulden and McIntosh 2018; and references therein). For climate change, studies have assessed historical trends (Nandintsetseg, Greene, and Goulden 2007; Vandandorj et al. 2017); initial projections of future change (Namkhaijantsan 2006); impacts linked with permafrost, livestock grazing, vegetation, and fires (e.g., Spence et al. 2014; Sharkhuu et al. 2016; Gradel et al. 2017); and community perspectives (Goulden et al. 2016). Civil society organizations (CSOs) (e.g., the Mongol Ecology Center [MEC]; the ecoLeap Foundation) and development agencies (e.g., the European Commission, KfW, the Swiss Agency for Development Cooperation, and the World Bank) have played a critical role in providing support for park management and livelihoods, including training and equipment for rangers, water quality sampling, waste management, and guidelines for the management of tourism, waste, and infrastructure (e.g., MEC 2013; Goulden and McIntosh 2018; ecoLeap Foundation 2020).

The KLNP is located within a biodiversity-rich region that supports four other national-level protected areas nearby—three in Khuvsgul *aimag* and an adjacent national park in the Russian Federation (Figure 1); this has climate change implications for park management (Improving Habitat Connectivity, Chapter 6).

3 How Is the Climate Changing?

Key Messages

- The climate at Khuvsgul Lake National Park (KLNP) has changed significantly over the past 50–60 years. Air and lake water temperatures have increased. Rainfall is less frequent and is more concentrated in heavy storms.
- Modeling indicates that, by 2050, there will be substantial further changes in the climate. Mean air temperatures are projected to continue rising. Small increases in summer and winter precipitation may occur in some regions. Evaporation rates will increase. Overall, the climate is projected to be warmer and drier. These changes are likely to differ in areas because of the diverse topography and elevations in KLNP. Temperature increases are projected to be smallest in the low-lying elevations of the park (floodplains and hills east of Khuvsgul Lake) and largest in the mountains east and west of Khuvsgul Lake. The mountains in the far east will show the largest increase in precipitation.
- The projected changes in climatic conditions have significant implications for the conservation values and livelihoods of KLNP (Chapters 4 and 5).

Climate of Khuvsgul Lake National Park

The climate of the Khuvsgul Lake National Park (KLNP) region is characterized by cold, dry winters and mild, windy summers (Namkhaijantsan 2006). Mean monthly temperatures range from −3.5°C to 18.5°C in summer (mean 15°C to 20°C in July, the warmest month) and from −23.2°C to −2.9°C in winter (mean −30°C to −25°C in January, the coldest month). Mountain summits may be 15°C–20°C cooler than the valleys, with mean air temperatures above 10°C occurring for around 90 days a year in the mountains, compared with 90–110 days in the valleys (Nandintsetseg, Greene, and Goulden 2007). There is a high number of sunny days (mean 260) per year (Dagvadorj, Batjargal, and Natsagdorg 2014). Mean annual potential evapotranspiration (PET) ranges from 600 millimeters (mm) around Khuvsgul Lake to less than 350 mm in the mountains. Mean annual precipitation is low (257 mm in Khankh town and 313 mm in Khatgal town). About 85% of total precipitation (rain and/or snow) falls between April and September, 50%–60% of this in July and August. Snow contributes less than 20% of total annual precipitation, and the annual duration of snow cover lasts 120–150 days a year (Batima 2006). Northern winds dominate in summer and spring, while southwesterly winds dominate in winter and autumn (Namkhaijantsan 2006). Because of the moderating effects of permafrost, topography, and Khuvsgul Lake, winter begins and ends later in the north of KLNP than in the south (Namkhaijantsan 2006).

Observed Changes, 1963–2016

Meteorological data on air temperature, water temperature in Khuvsgul Lake, rainfall, and storm events were compiled from weather stations in Khatgal and Khankh towns for the periods 1963–2016 (Khatgal) and 1985–2016 (Khankh). Data were analyzed for trends and supplemented by stakeholder observations. In the following discussion, in accordance with standard use, the mean minimum and mean maximum annual temperatures refer to the mean monthly minimum temperature (i.e., the average of all the daily coldest temperatures recorded in a month) and the mean monthly maximum temperature (i.e., the average of all the daily highest temperatures in a month), averaged over a year, to derive the mean annual temperature.

Air temperature. The compiled data indicate that the following changes had occurred: (i) mean annual temperatures in Khatgal town had increased by 0.31°C/decade (insufficient data is available to determine annual trends for Khankh town); (ii) mean, minimum, and maximum temperatures for spring and summer in both towns had increased by over 0.3°C/decade, and minimum spring temperatures in Khankh town increased by over 0.8°C/ decade; (iii) maximum summer temperatures had increased by 0.41°C/decade in Khatgal town and by 1.14°C/decade in Khankh town; (iv) mean autumn temperatures in Khatgal had increased by 0.4°C/decade; and (v) mean winter temperatures in both towns had remained virtually unchanged (Table 1).

Table 1: Measured Changes in Air Temperature in Khatgal Town (1963–2016) and Khankh Town (1985–2016) in Khuvsgul Lake National Park

Temperature Parameter[a]	Khatgal Town		Khankh Town	
	Average (1963–2016) (°C)	Measured Change (°C/decade)	Average (1985–2016) (°C)	Measured Change (°C/decade)
Mean annual	(4.5)	0.31	(3.8)	...
Mean annual maximum	3.5	0.36	2.2	...
Mean annual minimum	(11.7)	0.41	(8.0)	...
Spring mean	(3.5)	0.44	(3.9)	0.65
Summer mean	10.9	0.33	10.3	0.46
Autumn mean	(4.3)	0.40		...
Winter mean	(21.4)	n/s	(18.5)	n/s
Spring maximum	4.8	0.54	1.2	0.51
Summer maximum	17.9	0.41	17.1	1.14
Spring minimum	(11.8)	0.56	(9.4)	0.82
Summer minimum	4.0	0.42	6.1	0.49

( ) = negative, ... = not available (insufficient data available for analysis), n/s = no significant trend.

[a] Inclusive months: spring = March–May, summer = June–August, autumn = September–November, and winter = December–February.

Source: Asian Development Bank. Derived from raw data provided by the National Agency for Meteorology, Hydrology and Environmental Monitoring.

Water temperature of Khuvsgul Lake. Lake temperatures are recorded by the National Agency for Meteorology, Hydrology and Environmental Monitoring at a depth of 0.1 m near Khatgal and Khankh towns. These data indicate that (i) at the southern end of Khuvsgul Lake, near Khatgal town, the annual maximum water temperature had increased from around 14°C in the 1960s to 18°C in 2016 (no comparable data is available for Khankh); and (ii) in Khatgal town, the spring thaw of ice on Khuvsgul Lake now begins earlier (before March) compared with the 1960s and 1970s. This trend is less clear at the northern end of the lake, near Khankh town, where, after 1980, the start of spring thaw began earlier; but since 2003, it has returned to similar dates from 1970 to 1979. There is insufficient data to determine the cause of this pattern. However, the close relationship between ice melt and air temperature suggests that spring air temperatures may have decreased after 2003.

Rainfall and thunderstorm events. Data on daily rainfall in Khatgal town are available for the period 1963–2012 (no comparable data are available for Khankh town). These data indicate that there were no statistically significant changes in trends in rainfall amount, maximum number of consecutive dry days or wet days, or number of thunderstorms, during this time. Yet over the same period, the number of thunderstorms per year in Khatgal town showed a stepwise doubling, from a mean number of 15 thunderstorms per year in 1963–1979 to 30 per year in 1980–2012 (Goulden et al. 2016).

Stakeholder perceptions. Residents, tour operators, and staff of government agencies surveyed for the study reported changes in weather patterns that were largely consistent with the trends identified from meteorological data. Of 700 households interviewed, 595 (85%) reported that rivers, streams, and springs were drying up, and 574 (82%) reported that drought and declining rainfall were becoming more common. Most households (85%) felt that heat and aridity are pressing environmental issues. A total of 350 households (50%) felt that the number of fires was increasing and presented a serious threat to communities, but only 140 households (20%) felt that floods or heavy rainstorms were serious threats to communities.

Of the 33 personnel interviewed from 25 agencies (comprising 15 local government agencies, 6 CSOs, and 4 tour companies), the following were reported as changes in weather patterns: increasing drought (12 interviewees), warming conditions (10), declining rainfall (7), more intense rainfall (6), and stronger winds (3). They also reported the following impacts attributed to these changes: the drying up of springs or rivers (15 interviewees), desertification (6), declining condition of pasture and vegetation (4), and the melting of a small glacier (the only glacier in KLNP, located in mountains north of Khuvsgul Lake).

Other studies. The upward trend in air temperatures since the 1960s, documented in this study, is consistent with the findings of Namkhaijantsan (2006), who analyzed meteorological data from the same weather stations in Khatgal and Khankh towns as well as data from Renchinlkhumbe *soum*. In contrast, the study findings that there had been no long-term, statistically significant change in precipitation and little change in mean winter temperatures differ from Namkhaijantsan (2006), who found that precipitation and the duration of winter at KLNP have increased since the 1960s. These differences may be due to the use of slightly different data sets and analytical methods. Other studies on KLNP, northern Mongolia, and nearby regions of the Russian Federation have documented increased drought (Nandintsetseg and Shinoda 2013), increased fires (Shvidenko and Schepaschenko 2013), and a decline in vegetation quality (e.g., Chu and Guo 2012; Bayarjargal et al. 2019), linked with the drying climate, although the impacts of climate change and overgrazing are unclear.

Projected Changes by 2050

Climate change projections for KLNP were derived from global climate models for the following parameters: air temperature, precipitation, PET, aridity–wetness index (AWI), and growing degree days (GDD) (see Glossary). The mean values for the 2 decades centered on 2050 (2041–2060) were compared against the mean of the modeled conditions for the period 1960–1990 (the climate "baseline") under two representative concentration pathway (RCP) scenarios (Appendix). For air temperature, the parameters calculated were annual and seasonal mean, maximum, and minimum temperatures. For precipitation, mean seasonal and annual precipitation were calculated.

Air temperature. As modeled, mean annual temperatures for KLNP for the period 1960–1990 ranged from –7.2°C to –5.6° C (overall mean –6.6°C), and the mean annual maximum temperature in most of KLNP was around 0°C, with cooler values in the mountains and warmer values in the valleys (Table 2).

In contrast, by 2050, (i) increases in mean annual temperatures are projected to occur across KLNP; (ii) mean annual maximum temperatures may rise by 1.0°C–1.4°C (summer temperatures in July, the warmest month, may rise by 2.0°C–3.0°C); (iii) mean annual minimum temperatures may rise by 2.4°C–2.9°C; and (iv) minimum annual temperatures may rise by 3.0°C–4.0°C in the mountains west of Khuvsgul Lake and by 5°C or more in the mountains east of the lake (Table 3, Figure 2). Changes are projected to be smallest in the low-lying elevations of the park (floodplains and hills east of Khuvsgul Lake) and highest in the mountains east and west of Khuvsgul Lake.

Growing degree days. The mean annual number of GDD for KLNP for the period 1960–1990 was modeled to be 1,074 (Table 2). By 2050, the mean annual number of GDD is projected to have increased by 36%–45% (Table 3, Figure 3). This result is consistent with the projected increase in mean annual minimum temperature in KLNP (Figure 3) and implies that, by 2050, there will be more days with conditions warm enough for plant growth.

Table 2: Modeled Baseline (1960–1990, WorldClim) for Air Temperature and Growing Degree Days in Khuvsgul Lake National Park

Location	Mean Annual Air Temperature (°C)	Mean Annual Maximum Air Temperature (°C)	Mean Annual Minimum Air Temperature (°C)	GDD
Management Zone				
Limited use	(6.2)	0.7	(13.1)	1,151
Special	(7.2)	(0.5)	(13.9)	961
Tourism	(5.7)	1.3	(12.7)	1,173
All zones	(6.6)	0.2	(13.4)	1,074
Other Locations				
Khuvsgul Lake	(5.6)	1.4	(12.6)	1,285
Khatgal and Khankh towns	(6.0)	0.9	(12.9)	1,211

( ) = negative, GDD = growing degree days.
Source: Asian Development Bank. Derived from the models of Hijmans (2015) and Hijmans et al. (2005). See Appendix for the methodology.

Table 3: Projected Increases in Temperature and Growing Degree Days at Khuvsgul Lake National Park by 2050, Based on the WorldClim (1960–1990) Baseline

RCP Scenario	Mean Annual Temperature (°C)	Mean Annual Maximum Temperature (°C)	Mean Annual Minimum Temperature (°C)	GDD
RCP4.5	2.4	1.0	2.4	367
RCP8.5	2.9	1.4	3.0	444
RCP8.5 – GF[a]	4.9	3.9	4.5	786
RCP8.5 – MI[a]	5.3	3.6	5.5	830

GDD = growing degree days, RCP = representative concentration pathway.
[a] Climate models of high-risk scenarios (details are provided in the Appendix).
Source: Asian Development Bank. Derived from the models of Hijmans (2015) and Hijmans et al. (2005). See Appendix for the methodology.

Precipitation. For the period 1960–1990, mean annual precipitation at KLNP was 383 mm, as modeled, but was highly variable (ranging from 305 mm to 431 mm) (Table 4).

In contrast, by 2050, a small increase (34–37 mm) in mean annual precipitation across KLNP is projected for all climate scenarios, including the two high-risk RCP models (GF and MI; Appendix), with higher values estimated for RCP8.5 than for RCP4.5. Projected changes in summer precipitation are small or variable for different locations in the park. No change is forecast for areas north of Khuvsgul Lake. A moderate increase of up to 25 mm is projected for forest and forest steppe areas east of the lake, and higher increases of 50 mm or more for the eastern region of the park. Projected changes in winter precipitation are variable across the park, comprising a small increase (5–10 mm) across most of the park, a moderate increase (up to 20 mm) for hilly areas east of Khuvsgul Lake, and a higher increase (25 mm or more) for mountains on the far-east side of the park (Table 5, Figure 4).

Potential evapotranspiration. Under the modeled baseline conditions, the mean annual PET at KLNP from 1960–1990 was 556 mm (Table 4). By 2050, this is projected to have increased substantially (to 72–84 mm) across KLNP, as will the standard deviation for PET (Table 5). These projections indicate the likelihood of increased seasonality and larger interannual variation between seasons (Figure 5).

Figure 2: Projected Mean Annual Maximum and Minimum Temperatures at Khuvsgul Lake National Park by 2050, Compared with the Baseline (1960–1990, WorldClim)

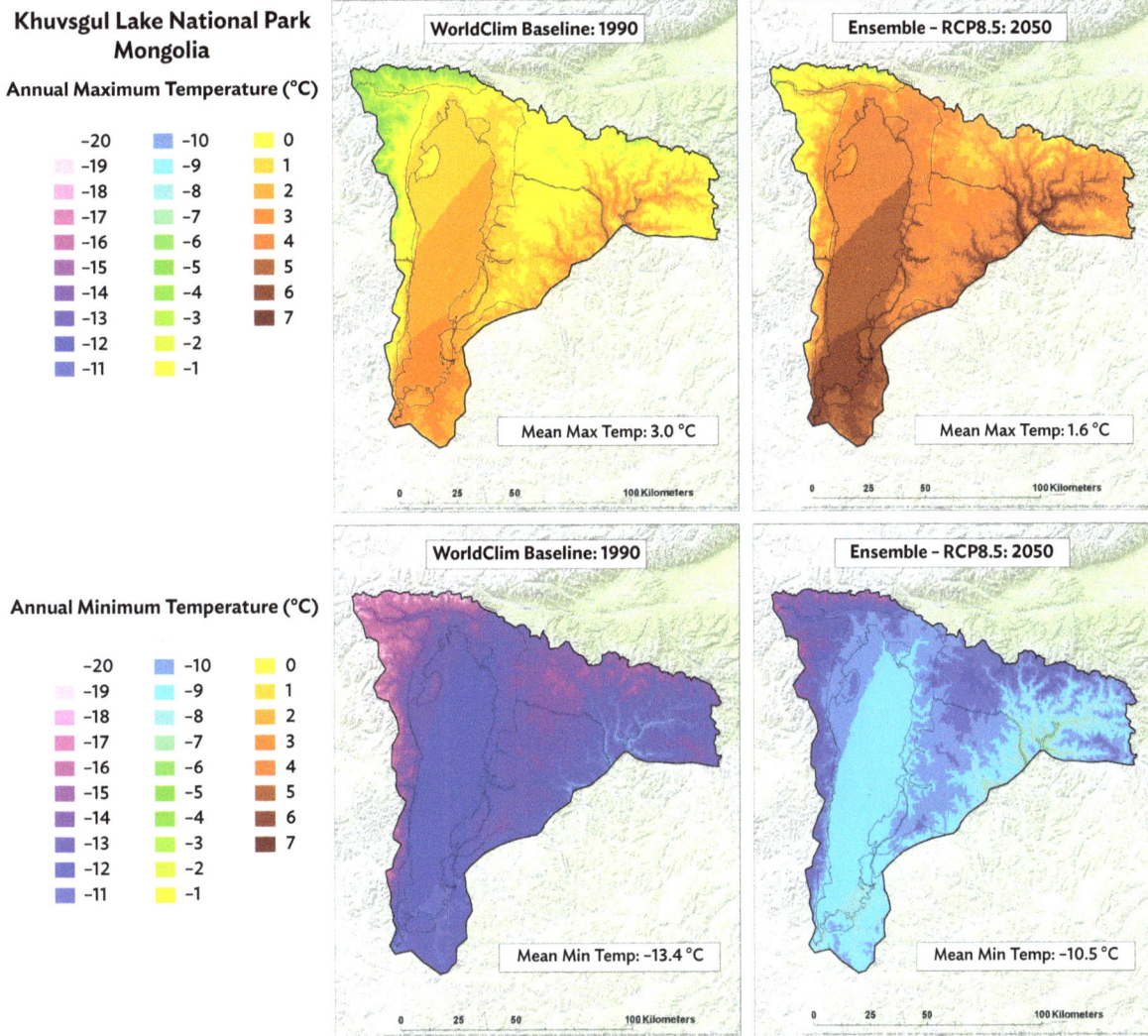

max = maximum, min = minimum, RCP = representative concentration pathway, temp = temperature.
Note: Based on RCP scenario 8.5 (RCP8.5).
Source: Asian Development Bank. Derived from the models of Hijmans (2015) and Hijmans et al. (2005). See Appendix for the methodology.

Aridity–wetness index. Under the modeled baseline for 1960–1990, the AWI for KLNP is estimated to have ranged from arid to semiarid (higher values represent wetter conditions), with a mean AWI of 0.7 (Table 4, Figure 6). This value is only slightly higher than the global threshold value of 0.65, which is generally taken to indicate adequate moisture for rainfed agriculture under semiarid conditions (Zomer et al. 2008). This implies that KLNP had a fairly dry climate. By 2050, the mean annual AWI for KLNP is projected to have decreased slightly, indicating a slight trend toward even drier conditions. This trend may be linked with the projected increase in PET, i.e., greater water loss due to evapotranspiration combined with drier and warmer conditions, which may, however, be partly offset by the projected small increases in precipitation in some regions of the park.

Figure 3: Projected Number of Growing Degree Days at Khuvsgul Lake National Park by 2050, Compared with the Baseline (1960–1990, WorldClim)

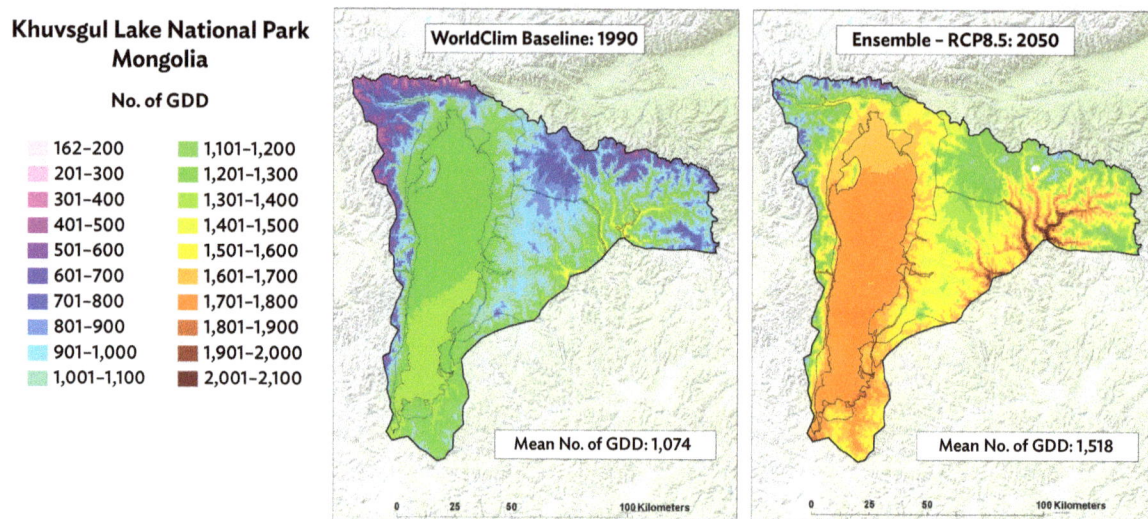

GDD = growing degree days, RCP = representative concentration pathway.
Note: Based on RCP scenario 8.5 (RCP8.5).
Source: Asian Development Bank. Derived from the models of Hijmans (2015) and Hijmans et al. (2005). See Appendix for the methodology.

Table 4: Modeled Baseline (1960–1990, WorldClim) for Precipitation-Related Parameters for Khuvsgul Lake National Park

Location	Mean Annual Precipitation	Mean PET	PET std	Mean AWI
	(mm)			
Management Zone				
Limited use	334	574	48	0.58
Special	431	532	45	0.82
Tourism	344	581	48	0.59
All zones	383	556	46	0.70
Other Locations				
Lake	305	600	49	0.51
Khatgal and Khankh towns	323	583	48	0.56

AWI = aridity–wetness index, mm = millimeter, PET = potential evapotranspiration, std = standard deviation.
Source: Asian Development Bank. Derived from the models of Hijmans (2015) and Hijmans et al. (2005). See Appendix for the methodology.

Other studies. Previous projections of air temperature and precipitation by 2050 at KLNP were made by Namkhaijantsan (2006), using data from the Khatgal meteorological station for the period 1963–1999. The study projected that, by 2050, the mean July air temperature would be 1.7°C higher and the mean summer precipitation would be 107.1 mm higher. For air temperature, this is less than the 2°C–3°C increase in the mean July temperature projected for 2050 under RCP8.5 in the current study. This indicates the rate of temperature increase is speeding up and is likely to result in further rapid changes. For summer precipitation, the increase

Table 5: Projected Changes in Precipitation-Related Parameters at Khuvsgul Lake National Park by 2050, Compared with the Baseline (1960–1990, WorldClim)

Location	Mean Annual Precipitation	Mean PET	PET std	Mean AWI
	(mm)			
RCP4.5	34	72	4	(0.02)
RCP8.5	37	84	5	(0.03)
RCP8.5 – GF	85	174	11	(0.05)
RCP8.5 – MI	48	140	7	(0.07)

( ) = negative, AWI = aridity-wetness index, mm = millimeter, PET = potential evapotranspiration, RCP = representative concentration pathway, std = standard deviation.

Source: Asian Development Bank. Derived from the models of Hijmans (2015) and Hijmans et al. (2005). See Appendix for the methodology.

Figure 4: Projected Mean Annual Precipitation in Khuvsgul Lake National Park by 2050, Compared with the Baseline (1960–1990, WorldClim)

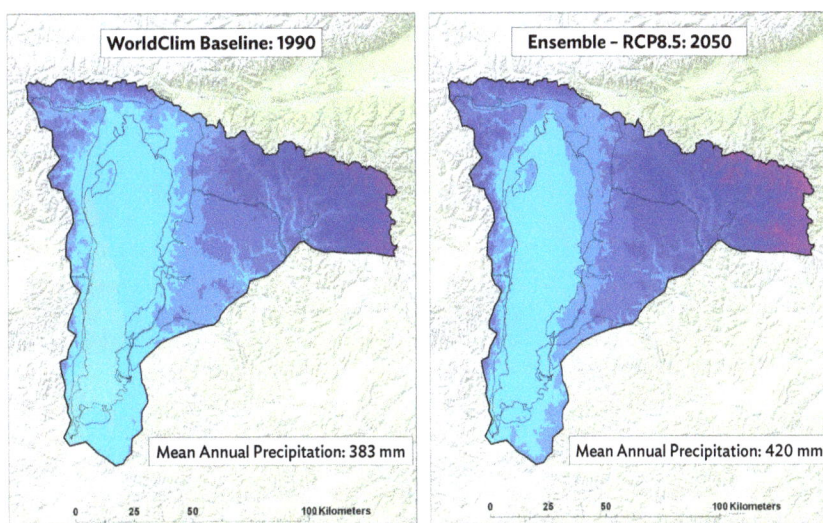

mm = millimeter, RCP = representative concentration pathway.

Note: Based on RCP scenario 8.5 (RCP8.5).

Source: Asian Development Bank. Derived from the models of Hijmans (2015) and Hijmans et al. (2005). See Appendix for the methodology.

projected in the current study (0–50 mm by 2050 under RCP8.5) is lower than the increase projected by Namkhaijantsan (2006); this difference is likely due to the use of different data sets.

The findings of the present study are broadly similar to those of two national-level modeling studies for Mongolia. The first study (Dagvadorj, Batjargal, and Natsagdorj 2014) projected increases in mean summer temperatures of 2.1°C by 2025 and 5.5°C by 2090, and a small increase in mean summer precipitation of 0%–5% by 2025 and 5%–10% by 2090, using a 10-model ensemble under the RCP8.5 scenario. The findings of the present study fall within these ranges. The second study (MET 2018) projected a 2°C increase in mean summer temperatures and an increase of 32 mm in mean annual evapotranspiration by 2055, and changes in precipitation of less than 5% by 2025 (RCP8.5, using regional models). In comparison, the present study projected slightly higher mean summer temperatures and a higher increase in evapotranspiration (72–84 mm), but made similar projections for precipitation (for the same time period and under RCP8.5).

Figure 5: Projected Mean Annual Potential Evapotranspiration at Khuvsgul Lake National Park by 2050, Compared with the Baseline (1960–1990, WorldClim)

Khuvsgul Lake National Park Mongolia

Potential Evapotranspiration (PET)

Annual PET (mm)

339–350	551–600
351–400	601–650
401–450	651–700
451–500	701–750
501–550	

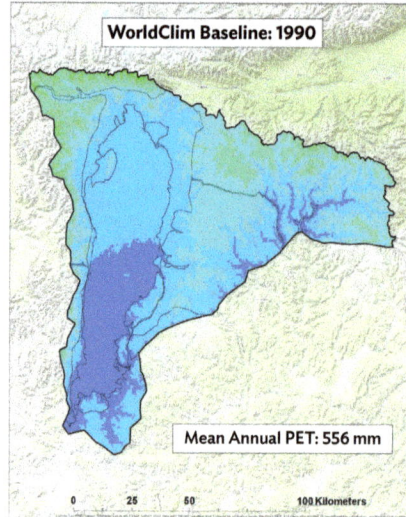

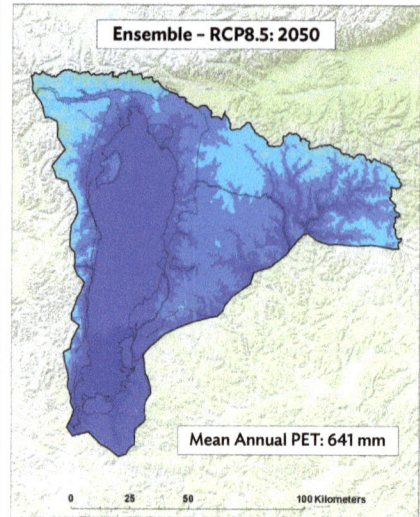

WorldClim Baseline: 1990

Mean Annual PET: 556 mm

Ensemble – RCP8.5: 2050

Mean Annual PET: 641 mm

mm = millimeter, RCP = representative concentration pathway.

Note: Based on RCP scenario 8.5 (RCP8.5).

Source: Asian Development Bank. Derived from the models of Hijmans (2015) and Hijmans et al. (2005). See Appendix for the methodology.

Figure 6: Projected Aridity–Wetness Index at Khuvsgul Lake National Park by 2050, Compared with the Baseline (1960–1990, WorldClim)

Khuvsgul Lake National Park Mongolia

Aridity–Wetness Index (AWI)

0.41–0.50	1.21–1.30
0.51–0.60	1.31–1.40
0.61–0.70	1.41–1.50
0.71–0.80	1.51–1.60
0.81–0.90	1.61–1.70
0.91–1.00	1.71–1.80
1.01–1.10	1.81–1.90
1.11–1.20	

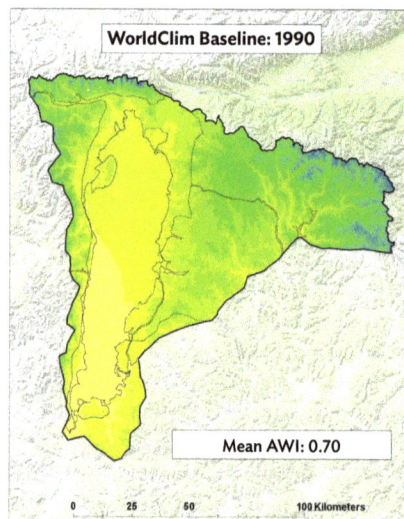

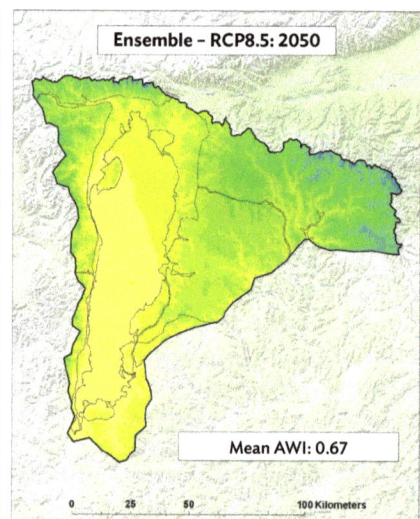

WorldClim Baseline: 1990

Mean AWI: 0.70

Ensemble – RCP8.5: 2050

Mean AWI: 0.67

RCP = representative concentration pathway.

Note: Based on RCP scenario 8.5 (RCP8.5).

Source: Asian Development Bank. Derived from the models of Hijmans (2015) and Hijmans et al. (2005). See Appendix for the methodology.

4 Bioclimatic Zones and Ecosystems

Key Messages

- Between 1960 and 1990, Khuvsgul Lake National Park (KLNP) was located within nine strata of three bioclimatic zones: extremely cold and wet zone, extremely cold and mesic zone, and cold and mesic zone. Most (58%) of KLNP was in the extremely cold and mesic zone.
- Based on an existing regional classification for northern Mongolia, 15 types of ecosystems occur in KLNP. These comprise 14 terrestrial (land-based) ecosystems (forest and steppe) and 1 aquatic ecosystem (Khuvsgul Lake).
- There is a close correlation between the distribution of ecosystems and bioclimatic zones in KLNP. This indicates that vegetation communities occur within, and depend upon, specific climatic and environmental conditions. These findings provide a baseline from which to assess the impacts of climate change (Chapter 5).

Bioclimatic Zones

A global environmental climate classification model that uses five climate parameters (Metzger et al. 2013) was applied to the baseline (1960–1990) climate of Khuvsgul Lake National Park (KLNP) (Chapter 3). The model categorizes the world into "bioclimatic zones," which represent the range of climate parameters that are most important for vegetation growth—e.g., temperature, aridity–wetness index (AWI), and growing degree days (GDD). The zones are further divided into "strata," representing finer-scale combinations of climate parameters. The terms "wet" (excess of moisture), "mesic" (less moist), "warm," and "cold" are used to compare bioclimatic zones within KLNP and the surrounding region. This approach resulted in a model representing the combined climatic and environmental conditions that plants and animals in KLNP lived within, under the baseline period of 1960–1990. This provided a basis to assess how climate change may affect the living conditions for vegetation, thereby impacting ecosystems. The methodology is discussed in the Appendix.

Baseline bioclimatic conditions. The KLNP had three bioclimatic zones during the baseline period: an extremely cold and wet zone (Zone D in Metzger et al. 2013, characterized by low mean temperatures, abundant moisture, and a short growing season); an extremely cold and mesic zone (Zone F, with adequate rather than abundant moisture, but a growing season similar to that in Zone D); and a cold and mesic zone (Zone G, with adequate moisture and a longer growing season). The extremely cold and wet zone, and the cold and mesic zone, each comprised a single stratum (coded as D_03 for the cold and wet zone, and G_01 for the cold and mesic zone), reflecting the relatively uniform bioclimatic conditions in these zones. The extremely cold and mesic zone consisted of seven strata (further details about the general characteristics of these strata can be found in the Appendix). The findings (Figure 7, Table 6) indicate the following baseline conditions for KLNP under each zone:

- The extremely cold and wet zone occupied high elevations in KLNP (mean 3,118 m), and had a mean annual temperature of less than –6.7°C, mean annual rainfall of 574 mm, mean annual AWI of 1.6, and mean GDD of 262 per year.
- The extremely cold and mesic zone occurred at middle elevations (2,000–3,000 m), and had a mean annual temperature between –4.3°C and 1.5°C, mean annual rainfall of 332–665 mm, and mean annual AWI of 0.6–1.6.
- The cold and mesic zone occurred at the lowest elevations in KLNP (mean 1,676 m), and had a mean annual temperature of 1.5°C, mean annual rainfall of 319 mm, mean annual AWI of 0.5, and mean GDD of 1,273 per year.

- Over 90% of KLNP was located within three strata of two zones: the extremely cold and mesic zone (strata F_08 and F_10) and the cold and mesic zone (G_01).
- The extremely cold and mesic zone encompassed the largest extent of the park (58% of the KLNP's total area).
- Among the strata, Stratum G_01 had the largest area (almost 5,000 km², or about 42% of the total). Most of this stratum was located at lower elevations, and included Khuvsgul Lake and river valleys.
- Stratum F_10 covered the second-largest area (almost 4,600 km², or about 39%).
- The extremely cold and wet zone, restricted to the highest elevations of the park, was the smallest in area (about 5 km²).

The findings indicate that the climate parameters of each bioclimatic zone showed consistent patterns across KLNP (Table 6):

- Mean annual temperatures were generally aligned and inversely correlated with mean elevation. Values ranged from less than

Figure 7: Bioclimatic Zones and Strata of Khuvsgul Lake National Park, under the Baseline Conditions (1960–1990, WorldClim)

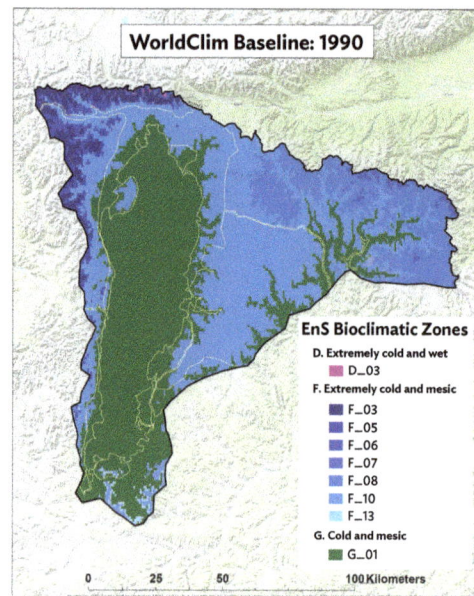

EnS = environmental strata.

Note: The Appendix provides details about the global model, zones, and strata relevant to this study.

Source: Asian Development Bank. Derived from the models of Hijmans (2015) and Hijmans et al. (2005).

Table 6: Bioclimatic Zones and Strata of Khuvsgul Lake National Park, under the Baseline Conditions (1960–1990, WorldClim)

Stratum[a]	Area (km²)	Mean Elev (masl)	MAT (°C)	Min Temp (°C)[b]	Max Temp (°C)[b]	MAP (mm)	Mean PET[b] (mm)	Mean AWI[b]	Mean GDD[b]
Extremely cold and wet (Zone D)									
D_03	5	3,118	(6.7)	(19.5)	(0.7)	574	367	1.6	262
Extremely cold and mesic (Zone F)									
F_03	397	2,590	(4.3)	(17.5)	(0.4)	446	443	1.0	523
F_05	345	2,351	(3.1)	(16.4)	(0.3)	398	480	0.8	684
F_06	131	2,597	(3.3)	(16.2)	(0.3)	609	444	1.4	553
F_07	4	2,795	(4.4)	(17.0)	(0.4)	665	412	1.6	432
F_08	1,329	2,207	(1.1)	(14.1)	(0.1)	541	505	1.1	835
F_10	4,589	1,971	(0.2)	(13.6)	0.0	394	545	0.7	1,021
F_13	64	1,966	1.5	(12.9)	0.1	332	577	0.6	1,117
Cold and mesic (Zone G)									
G_01	4,966	1,676	1.5	(12.5)	0.1	319	598	0.5	1,273

( ) = negative, AWI = aridity–wetness index, elev = elevation, GDD = growing degree days, km² = square kilometer, MAP = mean annual precipitation, masl = meters above sea level, MAT = mean annual temperature, max temp = maximum temperature, min temp = minimum temperature, mm = millimeter, PET = potential evaporation.

[a] The Appendix provides details about the global model, zones, and strata relevant to this study.

[b] Mean annual values.

Source: Asian Development Bank. Derived from the models of Hijmans (2015) and Hijmans et al. (2005).

−6.7°C for the coldest strata (mean elevation of 3,118 m) to 1.5°C for the warmest strata (mean elevation of less than 2,000 m).

- Annual precipitation generally increased with elevation, with lower elevation strata receiving the least precipitation.
- The AWI increased with increasing elevation, indicating wetter conditions, from the relatively dry stratum G_01 (value of 0.5) to the extremely cold and wet zone (1.6). The AWI values indicate that there were adequate moisture and temperature conditions conducive to vegetation growth, but these favorable conditions are likely to have been countered by the colder mean temperatures in higher elevation areas, which inhibit plant growth.
- The number of GDD also exhibited a pattern associated with elevation, with the highest elevations calculated to receive 262 GDD, and the lowest, 1,273 GDD.
- The KLNP special use zone (the largest management zone) had the most diverse climatic conditions, with nine strata, and was predominately (97%) composed of the two most extensive strata (F_10 and G_01).

Ecosystems

The ecosystems that occur within KLNP were identified from an existing regional ecosystem classification for northern Mongolia (Heiner et al. 2017). Of the 99 types of ecosystems in northern Mongolia (Heiner et al. 2017), a total of 15 occur in KLNP: 14 terrestrial (land-based) and 1 aquatic (Khuvsgul Lake) (Table 7, Figure 8). Four of the terrestrial ecosystems (riverine forest, riverine meadow, riverine shrub, and small riparian stream) may also be considered aquatic ecosystems, as they represent the distribution of freshwater physical habitat, based on biogeography (major basin), stream volume and annual flow (drainage catchment size), and geomorphology

Table 7: Ecosystems in Khuvsgul Lake National Park

Ecosystem Category	Area within KLNP			Area within Surrounding Region[a]		
	Area (km²)	% of KLNP	Mean Elevation (masl)	Area (km²)	% of KLNP	Mean Elevation (masl)
Alpine barren	36	0.3	2,932	117	0.2	2,848
Alpine steppe	905	7.7	2,432	6,753	10.3	2,443
Alpine tundra	169	1.4	2,695	1,812	2.8	2,645
Dry river	2	0.0	1,646	13	0.0	1,707
Forest closed	721	6.1	1,802	7,743	11.8	1,783
Forest open	1,974	16.7	1,881	13,213	20.2	1,789
Forest steppe	1,510	12.8	1,849	11,596	17.7	1,640
Lower mountain steppe	3,285	27.8	2,043	14,729	22.5	2,007
Meadow	145	1.2	1,714	1,166	1.8	1,587
Riverine forest	34	0.3	1,599	471	0.7	1,484
Riverine meadow	39	0.3	1,664	643	1.0	1,643
Riverine shrub	68	0.6	1,671	840	1.3	1,329
Small riparian stream	98	0.8	1,839	630	1.0	1,728
Steppe	42	0.4	1,734	2,767	4.2	1,786
Khuvsgul Lake	2,781	23.5	1,643	3,030	4.6	1,642
Total	**11,808**	**100.0**		**65,523**	**100.0**	

KLNP = Khuvsgul Lake National Park, km² = square kilometer, masl = meters above sea level.

Note: Numbers may not sum precisely and percentages may not total 100% because of rounding.

[a] Defined as the area of Khuvsgul *aimag* (province) north of latitude 49.8°N and west of 102.7°E. This encompasses all of the *soums* (districts) and protected areas around KLNP.

Source: Asian Development Bank. Derived from the ecosystem classification and data in Heiner et al. (2017).

Figure 8: Distribution of Ecosystems in Khuvsgul Lake National Park

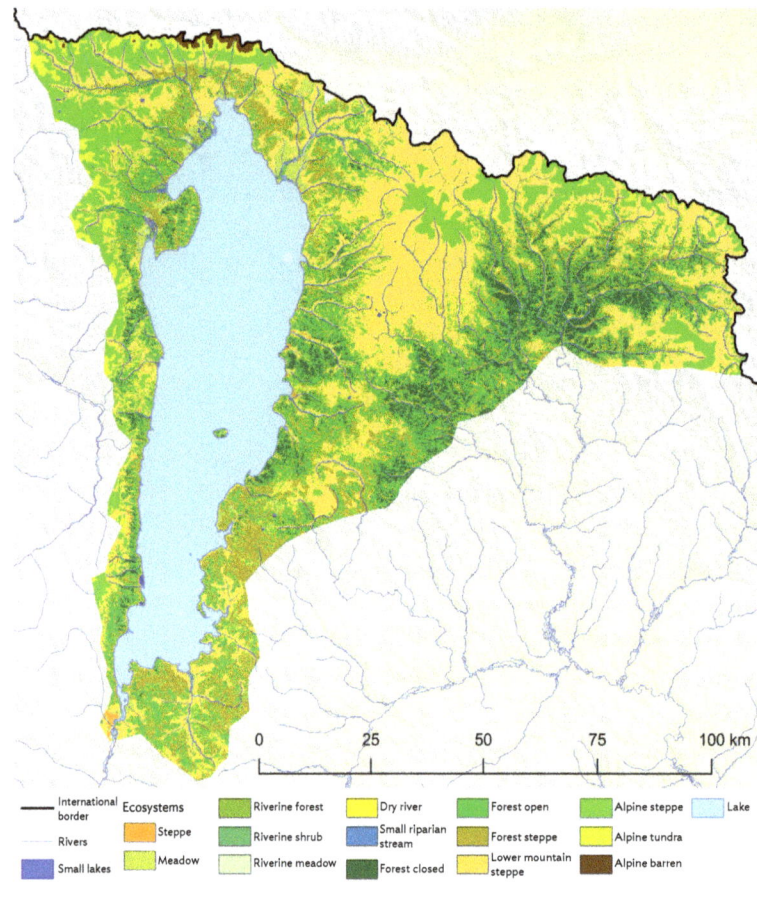

km = kilometer.

Note: Labeling for some of the categories is slightly different from the published names in Heiner et al. (2017).

Source: Asian Development Bank. Derived from the ecosystem classification and data in Heiner et al. (2017).

(Heiner et al. 2017). The classification indicates the following: (i) almost 81% of the park is occupied by four types of ecosystems (open forest, forest steppe, lower mountain steppe, and Khuvsgul Lake); (ii) after lower mountain steppe (27.8%), Khuvsgul Lake (23.5%) is the second-largest ecosystem in KLNP; and (iii) seven of the 15 ecosystems each occupy less than 1% of the park.

A limitation in the preparation of the regional ecosystem classification, described in Heiner et al. (2017), was the limited availability of detailed maps for vegetation, together with the inherent challenges of mapping over large areas with a wide range of data sets. Other regional studies have also used a range of scales and details to map vegetation. For the present study, to reduce the risk that spatial calculations based on the 15 ecosystem categories might present levels of accuracy higher than could be reasonably expected from the source data, the 15 categories were also aggregated into 7 categories (Table 8), which could be identified with confidence. The analysis of climate change impacts (Chapter 5) was applied to the original 15 categories and the 7 aggregated categories.

Table 8: Aggregated Ecosystem Categories for Khuvsgul Lake National Park

Aggregated Ecosystem Category	Area (km²)	% of KLNP	Ecosystem Categories in Heiner et al. (2017)
High mountain alpine	1,110	9.4	Alpine steppe, alpine tundra, alpine barren
Mountain steppe	3,528	29.8	Lower mountain steppe, meadow, small stream riparian
Forest steppe	1,510	12.8	Forest steppe
Taiga forest	2,729	23.1	Forest closed, forest open, riverine forest
Riparian areas	109	0.9	Riverine shrub, riverine meadow, dry riverbed
Steppe	42	0.4	Steppe (dry, moderately dry)
Lake	2,780	23.6	Water

KLNP = Khuvsgul Lake National Park, km² = square kilometer.
Source: Asian Development Bank. Derived from the ecosystem classification and data in Heiner et al. (2017).

These seven aggregated ecosystem categories are described briefly below.

High mountain alpine ecosystem (>2,300 meters elevation). This category comprises the mountain summits and upper slopes, along the north, west, and northeast boundaries of KLNP. Conditions are harsh, cold, windy, and frozen most of the year. Habitats include bare and rocky slopes, alpine meadows, and wet depressions, with thin soils, sedges, mosses, grasses, and shrubs. Few species of plants and animals occur. Wildlife includes herbivores—e.g., argali (*Ovis ammon*); Siberian ibex (*Capra sibirica*); reindeer (*Rangifer tarandus*); two rodents, the large-eared vole (*Alticola macrotis*) and the Tuva silver vole (*A. tuvinicus*); and a threatened bird, Altai snowcock (*Tetraogallus altaicus*).

Mountain steppe ecosystem (1,700–2,300 meters). This category comprises mountain slopes vegetated with grasses, sedges, shrubs, and low, scattered trees. This is the largest ecosystem in KLNP (Table 8). Small trees of larch (*Larix* spp.), birch (*Betula* spp.), and willow (*Salix* spp.) occur. The ecosystem supports rare plants and herbivores, e.g., reindeer and moose (*Alces alces*), and is used for livestock grazing and winter fodder (hay). The presence of small trees indicates that much of the ecosystem may originally have been forest, which has been lost as a result of cutting and overgrazing by livestock.

Forest steppe ecosystem (1,700–2,000 meters). This category comprises forests of larch and open steppe, on hills and lower mountain slopes in the northeast, southeast, and southwest of KLNP. The ecosystem partly overlaps with the mountain steppe ecosystem and is distinguished by Heiner et al. (2017) using the 1,950 m contour. The lower elevation of the ecosystem extends almost to Khuvsgul Lake. The forested areas support an understory of sedges, grasses, and rushes. The steppe areas support grasses, herbs, and other small, nonwoody plants. Similar to the mountain steppe, the ecosystem supports rare plants and herbivores (e.g., reindeer, moose) and is used for livestock grazing and fodder production.

Taiga forest ecosystem (1,645–2,400 meters). This category comprises mixed forests of coniferous and deciduous tree species, with an understory of shrubs, herbs, sedges, and grasses. Tree species include larch, birch, willow, Siberian pine (*Pinus sibirica*), spruce (*Picea obovata*), and aspen (*Populus tremula*). The ecosystem supports a wide range of fauna, including brown bear (*Ursus arctos*), wolverine (*Gulo gulo*), and deer species.

Riparian area ecosystem (1,500–1,650 meters). This category comprises a linear pattern of wet meadows and shrubs along streams that drain into Khuvsgul Lake. Soils are alluvial and have high organic matter and moisture content. Vegetation cover is dense and includes reeds, sedges, rushes, grasses, herbs, and shrubs, surrounded by small stands of willow, birch, larch, and other trees. These habitats provide important grazing and foraging areas for native animals, including moose, elk, deer, and wetland and grassland bird species. The ecosystem is used for pasture throughout the year, and large areas are overgrazed and degraded.

Steppe ecosystem (1,500–1,700 meters). This category comprises grasslands and meadows of grasses, sedges, and herbs on south-facing (sunny) mountain slopes and lands around Khuvsgul Lake. The ecosystem is used for pasture throughout the year, and large areas are overgrazed and degraded.

Khuvsgul Lake ecosystem (1,645 meters). Khuvsgul Lake has a volume of 380.7 cubic kilometers and a maximum depth of 262.4 m (Goulden et al. 2006). Large areas of the lake have a relatively uniform habitat of deep water (over 50 m), with a silty or rocky bed having few or no aquatic plants. Water circulation is slow, and nutrient levels and rates of bacterial decomposition are low. Much of the geology in the lake's basin is sedimentary, and this has a strong influence on the water chemistry by producing carbonate-rich alkaline water (Urabe et al. 2006). The water is mostly clear and of high quality. Exposed sections of shore are subject to strong wave action and have large rocks and boulders, and sheltered shores in small bays have rocky beaches of pebbles or coarse sand. Numerous small, semi-enclosed bays occur around the lake, and these have limited water mixing with the lake; this water mixing has important implications for water quality under climate change (Khuvsgul Lake, under Ecological Values, Chapter 5). Small estuaries occur along the eastern, southern, and northern shorelines. These contain nutrients from vegetation (resulting in acidic, brown-colored water), with higher levels of nitrogen (Urabe et al. 2006) and daily temperature fluctuations (Hayami et al. 2006) than in deeper parts of the lake.

Comparison with surrounding regions

To understand the extent to which the 15 ecosystem categories in KLNP are represented in the wider region—an important consideration for conservation planning, especially under climate change—the presence of each category outside the park was assessed, through two measures: (i) the area of each ecosystem up to 50 km outside KLNP (measured in 10 km increments); and (ii) the area of each ecosystem within all lands of the four *soums* where KLNP is located, excluding the park area. The first measure encompassed portions of two nearby protected areas (Figure 1), the Khoridol Saridag Strictly Protected Area (of which 2,107 km², or 93%, was encompassed within the 50 km search area) and the Tengis–Shishged National Park (1,431 km², or 17%). Adjacent lands in the Russian Federation were also within the 50 km search area, but these were excluded from analysis on account of a lack of comparable mapping data on ecosystems.

For the measure using a 50 km search area, the key findings (Table 9) were as follows: (i) the area of open water outside KLNP is much smaller, because of the absence of large lakes comparable to Khuvsgul Lake; (ii) except for Khuvsgul Lake, most of the 14 terrestrial ecosystem categories are relatively well represented outside KLNP, including those within 20 km of the park boundaries; (iii) similar to KLNP, the five largest ecosystem categories outside KLNP are lower mountain steppe, open forest, closed forest, forest steppe, and alpine steppe; (iv) alpine barren is the smallest of all categories—KLNP supports a notably higher area of the alpine barren category than any other lands within 50 km in Khuvsgul *aimag*; (v) outside KLNP, two ecosystem categories are found at elevations lower than those within KLNP, open forest (mean elevation almost 100 m lower) and forest steppe (over 200 m lower); (vi) two other ecosystem categories, closed forest and alpine steppe, each make up more than 10% of the surrounding area and occur at elevations similar to those within KLNP; (vii) the riverine shrub ecosystem is found at much lower elevations in surrounding areas; and (viii) the remaining categories are generally limited in extent within and outside KLNP. These patterns partly reflect a larger range of elevations outside KLNP (higher peaks and lower valleys), although the mean elevation of ecosystems is generally similar to that within KLNP.

For the measure comparing the spatial extent of the ecosystem categories within surrounding *soum* lands, the key findings (Table 10) were as follows: (i) of the four *soums*, Renchinlkhumbe *soum* supports the largest areas of all 14 terrestrial ecosystems outside KLNP, except for three (open forest, forest steppe, and riverine shrub), for which Tsagaan–Üür *soum* has the largest areas; and (ii) for three ecosystems (alpine barren, alpine tundra, and dry river), Renchinlkhumbe *soum* is the only one of the four *soums* that supports these categories outside KLNP.

These findings have implications for park management, and these are discussed in Chapter 6.

Table 9: Spatial Extent of the Ecosystem Categories of Khuvsgul Lake National Park within a 50-kilometer Search Area, including Two Protected Areas
(km²)

Ecosystem Category[a]	KLNP	Search Distance outside KLNP				Protected Area Encompassed within 50 km Search Area	
		20 km	30 km	40 km	50 km	Khoridol Saridag SPA	Tengis-Shishged National Park
Alpine barren	36	2	2	6	21	0	19
Alpine steppe	905	1,118	1,565	1,997	2,254	985	345
Alpine tundra	169	314	402	552	611	251	163
Dry river	2	4	6	7	10	2	0
Forest closed	721	802	1,227	1,515	1,763	91	194
Forest open	1,974	1,776	2,596	3,421	4,172	212	275
Forest steppe	1,510	1,709	2,818	3,926	5,058	39	6
Lower mountain steppe	3,285	1,785	2,460	3,278	4,421	484	387
Meadow	145	32	210	436	635	2	5
Riverine forest	34	36	95	121	169	1	14
Riverine meadow	39	73	138	214	320	17	5
Riverine shrub	68	65	189	260	350	1	4
Small riparian stream	98	79	119	163	210	22	17
Steppe	42	101	194	428	776	8	4
Total[b]	**11,808**	**7,896**	**12,022**	**16,323**	**20,770**	**2,113**	**1,440**

KLNP = Khuvsgul Lake National Park, km = kilometer, km² = square kilometer, SPA = strictly protected area.

[a] Excludes one category, Khuvsgul Lake, which occurs only in KLNP.
[b] Total figures were calculated inclusive of Khuvsgul Lake, which has an area of 2,781 km². Numbers may also not sum precisely because of rounding.

Source: Asian Development Bank.

Table 10: Spatial Extent of the Ecosystem Categories of Khuvsgul Lake National Park within Surrounding *Soum* Lands
(km²)

Ecosystem Category[a]	Within KLNP	Outside KLNP[b]			
		Alag–Erdene	Chandmani–Undur	Renchinlkhumbe	Tsagaan–Üür
Alpine barren	36	0	0	3	0
Alpine steppe	905	44	17	1,367	11
Alpine tundra	169	0	0	376	0
Dry river	2	0	0	8	0
Forest closed	721	69	63	1,066	638
Forest open	1,974	666	884	989	1,700
Forest steppe	1,510	623	2,198	164	2,928
Lower mountain steppe	3,285	1,663	338	1,521	203
Meadow	145	15	7	698	8
Riverine forest	34	3	7	142	38
Riverine meadow	39	101	6	239	6
Riverine shrub	68	21	80	124	188
Small riparian stream	98	39	36	70	53
Steppe	42	475	1	368	0
Total[c]	**11,808**	**3,719**	**3,636**	**7,137**	**5,774**

KLNP = Khuvsgul Lake National Park, km² = square kilometer.

[a] Excludes one category, Khuvsgul Lake, which occurs only in KLNP.
[b] Defined as the net area of lands outside KLNP of the four *soums* (districts) in which KLNP is located.
[c] Total figures were calculated inclusive of the Khuvsgul Lake, which has an area of 2,781 km². Numbers may also not sum precisely because of rounding.

Source: Asian Development Bank. Derived from ALAGaC (unpublished) and MET (2019).

Matching Bioclimatic Zones with Ecosystems

The three bioclimatic zones and their strata derived for the baseline climate (1960–1990) of KLNP were overlaid on the 15 ecosystem categories in Heiner et al. (2017) (Figure 9), and the extent of each ecosystem within the zones and strata was quantified (Table 11). The key findings were as follows:

- There is a relatively close (but less than 100%) correspondence between the bioclimatic zones and categories of ecosystems.
- Four strata (D_03, F_03, F_06, and F_07) are occupied by a single ecosystem category (alpine); most of F_05 (94%) and F_08 (91%) are occupied by one of the steppe-type ecosystem categories; and F_10 (88%) and G_01 (89% of the terrestrial area) are occupied by one of the categories made up of a mix of

Figure 9: Bioclimatic Zones Overlaid with the Ecosystem Categories in Heiner et al. (2017) for Khuvsgul Lake National Park, under the Baseline Conditions (1960–1990)

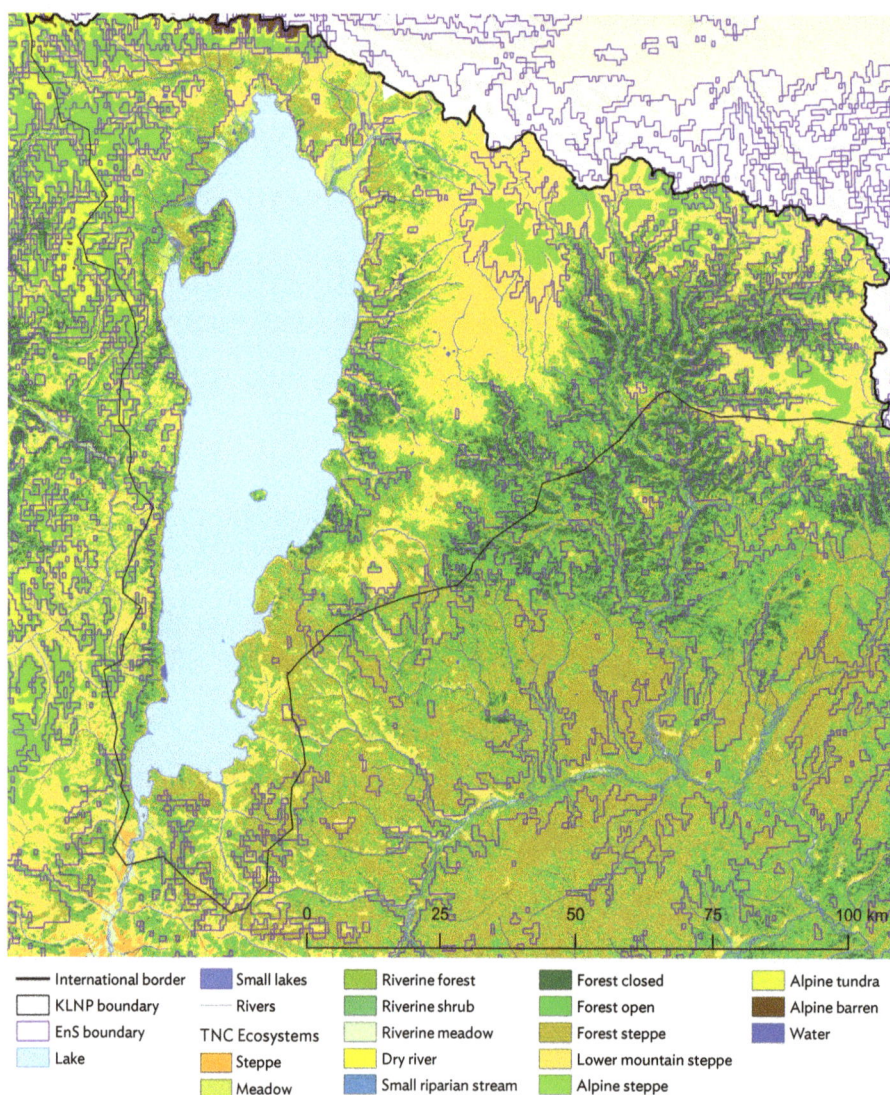

EnS = environmental strata, KLNP = Khuvsgul Lake National Park, TNC = The Nature Conservancy.
Note: The bioclimatic zones identified in this study are presented in the figure as purple polygons.
Source: Asian Development Bank. Derived from the ecosystem classification and data in Heiner et al. (2017).

Table 11: Proportion of Area Occupied by Ecosystems under Environmental Strata in Khuvsgul Lake National Park, under the Baseline Conditions (1960–1990)
(%)

Ecosystem Category	D_03	F_03	F_05	F_06	F_07	F_08	F_10	F_13	G_01
Alpine barren	13.3	30.9	0	53.8	2.0	0	0	0	0
Alpine steppe	0	27.4	23.7	8.4	0.2	36.1	4.1	0.2	0
Alpine tundra	0.1	76.0	6.5	13.9	0.9	1.2	1.4	0	0
Dry river	0	0	0	0	0	0	0	0	100.0
Forest closed	0	0	0	0	0	2.0	57.4	0	40.5
Forest open	0	0.1	0.3	0	0	5.0	64.9	1.1	28.7
Forest steppe	0	0	0	0	0	2.6	55.2	1.8	40.4
Lower mountain steppe	0	0.2	3.3	0.2	0	25.5	57.5	0.4	12.8
Meadow	0	0	0	0	0	0	24.6	0	75.4
Riverine forest	0	0	0	0	0	0	17.2	0	82.8
Riverine meadow	0	0	0	0	0	0	6.1	0	93.9
Riverine shrub	0	0	0	0	0	0	21.0	0	79.0
Small riparian stream	0	0.3	3.0	0	0	6.8	50.3	0	39.6
Steppe	0	0	0	0	0	1.4	22.0	0	76.6
Khuvsgul Lake	0	0	0	0	0	0	0.1	0	99.8

D = extremely cold and wet zone, F = extremely cold and mesic zone, G = cold and mesic zone.
Notes:
1. Ecosystem categories are derived from Heiner et al. (2017).
2. Climate codes are derived from Metzger et al. (2013).
3. The Appendix provides details about the global model, zones, and strata relevant to this study.
Source: Asian Development Bank.

forest and steppe (open forest, forest steppe, and lower mountain steppe, but excluding steppe); and 77% of F_13 is occupied by forest ecosystems, and the rest by lower mountain steppe.

- Twelve of the 15 ecosystems occur mainly in two strata (F_10 and G_01), and five occur only within these two strata.
- Large areas of four ecosystems (lower mountain steppe, forest, riverine, and steppe) occur within one stratum (G_01).
- Most (about 58%) of one ecosystem, the lower mountain steppe (the largest ecosystem in KLNP), occurs within one stratum (F_10).
- Three categories of forest ecosystems, which compose over 35% of KLNP, are located mainly in two strata (F_10, >55% of each forest type; and G_01, 30%–40%).

These findings indicate that under the baseline climate between 1960 and 1990, the ecosystems of KLNP occurred within a relatively low diversity of bioclimatic conditions across much of the park.

5 Impacts of Climate Change

Key Messages

- By 2050, Khuvsgul Lake National Park (KLNP) is projected to have undergone a profound change in bioclimatic conditions. The coldest bioclimatic zone in the park may disappear, and about 10,983 km^2 (93%) of the park is projected to become warmer and drier.
- These changes will almost certainly cause severe and irreversible impacts on biodiversity. Ecosystems at lower elevations are projected to expand. Small alpine ecosystems will decline or be lost. At least one rare, high-altitude plant species may become locally extinct. Ecosystems with dense, wet vegetation (closed forest, meadow, and riverine ecosystems) will decline and be replaced by drier, open forest and steppe, reducing the thick cover and rich feeding resources for large grazing mammals. Climate change will also compound existing threats to biodiversity. Plant and animal populations will be forced to adapt to new climate conditions or—if there is sufficient habitat and no barriers to movement—shift northward or to higher elevations to remain within cooler and wetter conditions. Overall, large changes in the composition of plant and animal communities in KLNP are likely to occur.
- For Khuvsgul Lake, warmer waters and increased nutrient inputs due to larger storms and winds, combined with unmanaged livestock waste and sewage, are likely to cause widespread water pollution. These changes would impact the unique communities of aquatic invertebrates and fish in the lake, which depend on high water quality and existing temperature regimes. This may place further pressure on fish species already threatened by overfishing, e.g., the unique Khuvsgul grayling. Reduced outflows from the lake due to the drying climate would also affect water supply across the lake's watershed, which extends across northern Mongolia.
- For livelihoods, the areas of open forest and steppe utilized for livestock grazing may increase, but this is unlikely to benefit livelihoods, as increasing drought, storms, and severe winters would compound the existing impacts of overgrazing. Poor herding households are likely to be the most affected by climate change. Hunting and fishing may increase, placing further pressures on biodiversity.
- For tourism, climate hazards may affect the integrity of infrastructure, increase operational costs, affect visitor safety, and reduce the visitor experience. A decline in tourism would reduce the opportunity for income diversification, a key measure to build livelihood resilience to climate change for local communities.
- The effects of climate change and existing threats are interlinked and require new approaches for the management of KLNP. This matter is discussed in Chapter 6.

Bioclimatic Zones

The modeled changes in climate at Khuvsgul Lake National Park (KLNP) that are projected to occur by 2050 under the RCP8.5 scenario (Chapter 3) were compared with the bioclimatic zones and ecosystem categories of KLNP under the baseline climate (1960–1990) (Chapter 4). The projected changes in distribution and area of the bioclimatic zones and their strata by 2050 were calculated (Appendix). The results (Figure 10, Table 12) indicate that climate change will cause large shifts in the bioclimatic zones of KLNP by 2050, as follows:

- By 2050, about 93% (10,983 km^2) of KLNP will have shifted to an entirely different set of bioclimatic conditions not previously experienced in that location.
- This shift is projected to include the entry of new bioclimatic zones into the park.
- Only about 7% (846 km^2) of KLNP will remain in its baseline bioclimatic zones.

Figure 10: Bioclimatic Zones of Khuvsgul Lake National Park (under the Baseline Conditions) and Projected Changes by 2050 (under RCP8.5)

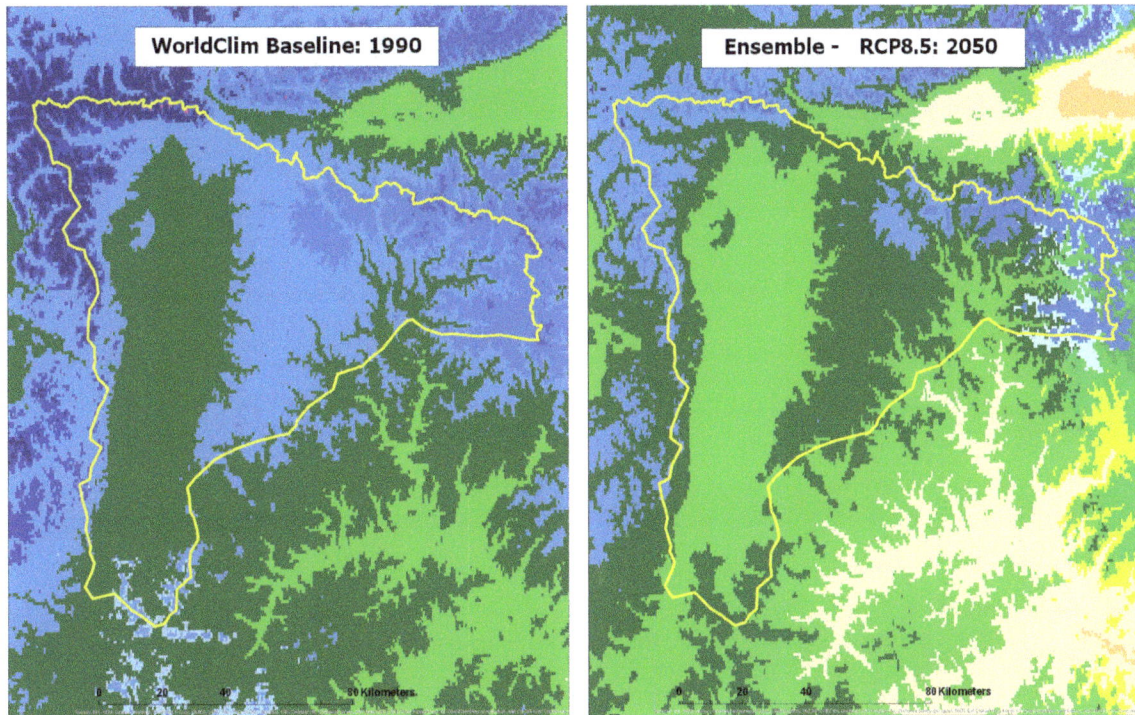

EnS Categories

D_03	F_06	F_12	G_01	G_07	H_01
F_03	F_07	F_13	G_02	G_08	H_03
F_04	F_08	F_14	G_04	G_09	H_04
F_05	F_10	F_15	G_05	G_10	H_06
	F_11		G_06	G_12	

D = extremely cold and wet zone, EnS = environmental strata, F = extremely cold and mesic zone, G = cold and mesic zone, H = cool temperate and dry zone, RCP = representative concentration pathway.

Notes:
1. Climate codes are derived from Metzger et al. (2013).
2. The Appendix provides details about the global model, zones, and strata relevant to this study.

Source: Asian Development Bank.

- The area of the park currently within the extremely cold and mesic zone (Zone F) will diminish significantly, from 58% of the park's area in 1960–1990 to 16% by 2050.
- The mean elevations of all bioclimatic zones in KNLP are projected to shift upward by 2050. Much of the extremely cold and wet zone will shift upward by 200–400 m and be replaced by the cold and mesic zone.
- The area of KLNP encompassed within the cold and mesic zone will almost double: over 82% of KLNP will eventually be in this zone, compared with 42% under baseline conditions. Conditions almost throughout the park will become warmer.
- Khuvsgul Lake will eventually be within warmer environmental conditions that currently do not occur within KLNP.

At the finer level of environmental strata within each bioclimatic zone, the findings (Table 12) indicate that, by 2050, (i) a new stratum (G_06) of the cold and mesic zone that did not occur in KLNP between 1960 and 1990 will eventually occupy almost 42% (4,923 km²) of KLNP. This will have higher annual temperatures than the existing

Table 12: Projected Changes in the Distribution of Environmental Strata in Khuvsgul Lake National Park by 2050, Compared with the Baseline Conditions (1960–1990)
(km²)

	Code	Area in 1990	EnS – 2050 Projection										
			F_03	F_06	F_07	F_08	F_10	F_14	F_15	G_01	G_04	G_05	G_06
			New Area in 2050										
	D_03	5	0	5	0	0	0	0	0	0	0	0	0
	F_03	397	2	0	0	78	316	0	0	<1	0	0	0
	F_05	345	0	0	0	1	328	0	0	16	0	0	0
	F_06	131	1	22	0	107	0	0	0	0	0	0	0
EnS – 1990	F_07	4	0	3	1	0	0	0	0	0	0	0	0
Baseline	F_08	1,329	0	0	0	516	341	12	138	265	36	22	0
	F_10	4,589	0	0	0	0	14	0	2	4,211	0	103	258
	F_13	64	0	0	0	0	2	0	0	63	0	0	0
	G_01	4,966	0	0	0	0	0		2	290	0	9	4,665
	Total		3	30	1	702	1,001	12	142	4,845	36	134	4,923

D = extremely cold and wet zone, EnS = environmental strata, F = extremely cold and mesic zone, G = cold and mesic zone, km² = square kilometer.

Notes:
1. Climate codes are derived from Metzger et al. (2013).
2. The Appendix provides details about the global model, zones, and strata relevant to this study.

Source: Asian Development Bank.

G_01 stratum (mean 1.4°C warmer and maximum 1.6°C warmer), a higher potential evapotranspiration (PET) (64 mm more than G_01 in 2050), and a lower mean precipitation (80 mm less). The projected aridity–wetness index (AWI) will fall below a global threshold value (0.65, Chapter 3) recognized as the limit for rainfed agriculture, indicating that ecosystems in this stratum will experience moisture stress in most years; (ii) most of the area within stratum G_01 will shift to the new stratum G_06, and most of the area within stratum F_10 (occupying 39% of KLNP under the baseline climate) will shift to stratum G_01; and (iii) one zone (extremely cold and wet zone) and three strata of the extremely cold and mesic zone (F_05, F_07, and F_13) are projected to disappear completely from KLNP. Although this zone and strata represent a small area of KLNP, they support restricted environmental and climatic conditions utilized by a small number of plant and animal species.

Projected changes by 2050 in the elevation of environmental strata within each bioclimatic zone, under RCP8.5 scenario (Figure 11, Table 13) indicate that the mean elevation and minimum and maximum elevations of most strata will shift upward, on average by 270 m. Two strata, which encompassed the largest regions of KLNP under the baseline climate (1960–1990), exhibit the greatest projected changes in elevation: stratum F_10 (an upward shift of 420 m) and G_01 (310 m). In comparison, modeling using the RCP4.5 scenario (not shown) resulted in an upward shift in the mean, maximum, and minimum elevations of each strata by 250 m.

High-risk scenarios. The projected impacts of climate change under two high-risk climate models (GF and MI; Appendix) on the elevations of bioclimatic zones are shown in Table 13.

The two high-risk climate models predict larger changes compared with the results of the other models used for the study (Appendix): (i) under one high-risk model, one additional stratum (F_03) disappears entirely from KLNP; (ii) under both models, two new strata (F_11 and G_10) appear in the park; and (iii) under one model, a third new stratum (G_02) appears in the park. These results emphasize the potentially extreme changes in climatic conditions that may occur in KLNP and a dramatic shift from extremely cold and wet conditions to warmer and drier conditions.

Figure 11: Projected Changes in the Elevations of Environmental Strata in Khuvsgul Lake National Park by 2050, Compared with the Baseline Conditions (1960–1990)

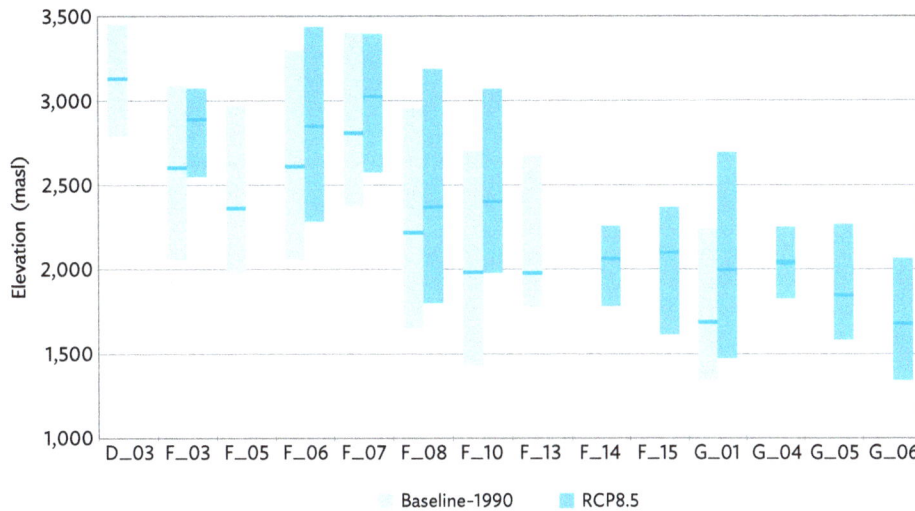

D = extremely cold and wet zone, F = extremely cold and mesic zone, G = cold and mesic zone, masl = meters above sea level, RCP = representative concentration pathway.

Notes:
1. Climate codes are derived from Metzger et al. (2013).
2. The Appendix provides details about the global model, zones, and strata relevant to this study.

Source: Asian Development Bank.

Table 13: Mean Elevation of Bioclimatic Zones of Khuvsgul Lake National Park (under the Baseline Conditions) and Projected Changes by 2050 (under RCP8.5) (m)

Zone/Stratum	Baseline (1960–1990)	Mean Elevation of Environmental Stratum			Upward Shift		
		Ensemble (10 models) RCP8.5	High-Risk Model GF	High-Risk Model MI	Ensemble (10 models) RCP8.5	High-Risk Model GF	High-Risk Model MI
Extremely cold and wet							
D_03	3,118	a	a	a	a	a	a
Extremely cold and mesic							
F_03	2,590	2,876	a	a	286	a	a
F_05	2,351	a	a	a	a	a	a
F_06	2,597	2,837	2,977	3,083	240	379	486
F_07	2,795	3,014	3,321	a	219	526	a
F_08	2,207	2,359	2,696	2,798	152	488	591
F_10	1,971	2,392	2,666	2,675	421	695	704
F_11	Not present	Not present	2,460	2,483	n/a	b	b
F_13	1,966	a	a	a	a	a	a
F_14	Not present	2,051	2,439	2,399	b	b	b
F_15	Not present	2,088	2,407	2,474	b	b	b

continued on next page

Table 13   *continued*

| Zone/Stratum | Mean Elevation of Environmental Stratum | | | | Upward Shift | | |
	Baseline (1960–1990)	Ensemble (10 models) RCP8.5	High-Risk Model GF	High-Risk Model MI	Ensemble (10 models) RCP8.5	High-Risk Model GF	High-Risk Model MI
Cold and mesic							
G_01	1,676	1,986	2,272	2,304	310	596	628
G_02	Not present	Not present	2,197		n/a	b	
G_04	Not present	2,025	2,192	2,245	b	b	b
G_05	Not present	1,837	1,968	2,046	b	b	b
G_06	Not present	1,670	1,800	1,925	b	b	b
G_07	Not present	Not present	1,663		b	b	
G_09	Not present	Not present	1,837		b	b	
G_10	Not present	Not present	1,667	1,779	n/a	b	b
Cool temperate and dry							
H_01	Not present	Not present	1,492	1,653	b	b	b

D = extremely cold and wet zone, F = extremely cold and mesic zone, G = cold and mesic zone, H = cool temperate and dry zone, m = meter, n/a = not applicable, RCP = representative concentration pathway.

Notes:

1. Climate codes are derived from Metzger et al. (2013).
2. The Appendix provides details about the global model, zones, and strata relevant to this study.

a No longer present in Khuvsgul Lake National Park by 2050.
b Did not occur during the baseline but projected to occur in the future.

Source: Asian Development Bank.

Ecological Values

Terrestrial Ecosystems

The large projected changes in the bioclimatic zones of KLNP imply profound impacts on the ecosystems and biodiversity values of KLNP. As temperatures, humidity, and other parameters change, the conditions required for plant growth will become more (or less) suitable for plant species and will, in turn, affect entire vegetation communities. Three major changes are likely to occur in ecosystems in KLNP: (i) the highest-elevation ecosystems in the park, which are the most adapted to cold and wet conditions, will decline or be lost entirely; (ii) the ecosystems that occur at lower elevations will expand to higher elevations; and (iii) there may be conversions of some ecosystems to other types of ecosystems (e.g., from closed forests to open forests). Plant and animal populations will either manage to adapt to the new climate conditions or shift northward or to higher elevations to remain within cooler and wetter conditions—assuming that they are sufficiently mobile, there is adequate habitat, and there are no barriers to movement—or failing these, decline or become locally extinct. Across all ecosystems, the composition of vegetation communities may disassemble and form novel communities, as species that favor warmer and drier conditions are able to colonize areas they previously did not occur in. Overall, large, irreversible changes in the composition of plant and animal communities in KLNP are likely to occur.

As the modeled climate baseline (1960–1990) was already 30 years old at the time of the study, and the climate projections are for 2050, it is likely that these changes are already well underway. These changes are further assessed as follows.

Loss or severe decline of ecosystems from Khuvsgul Lake National Park. Three ecosystems (alpine barren, alpine steppe, and alpine tundra) are projected to be the most highly impacted by climate change. By 2050, the environmental strata within which these ecosystems occur will have disappeared from KLNP or be drastically reduced in extent, and replaced by warmer and drier conditions. As a result, the alpine barren ecosystem may decline by 89%–92%, from about 36 km^2 to 3–4 km^2; alpine steppe may decline by 86%, from about 905 km^2 to 130 km^2; and alpine tundra may decline by 88%, from about 169 km^2 to 20 km^2 (Table 14). The alpine vegetation in these ecosystems may be replaced by grasses and shrubs of the mountain steppe vegetation, the ecosystem that occurs below the alpine zone (Chapter 4).

At least two mammal species, the large-eared vole and the Tuva silver vole (small rodents), and one plant species (*Saussurea involucrata*) may be impacted by the loss of alpine conditions. The two mammals occur in alpine and sub-alpine rocky fields, and little is known about their status in Mongolia (Clark et al. 2006). Their small size, limited mobility, and specialized reliance on alpine conditions indicate a low resilience to climate change. Compared with larger mammals for which other factors (e.g., hunting) are of more immediate threat, climate change may be the greatest threat to these species. *Saussurea involucrata* is a globally endangered, slow-growing species restricted to high-elevation rocky habitats (Chik et al. 2015). It is threatened because of over-collection for use in traditional medicine. In KLNP, *S. involucrata* occurs within a highly restricted area of less than 200 ha in the alpine tundra and alpine steppe ecosystems. The severe decline of both ecosystems, together with the impacts of collection, may cause the local extinction of the species in KLNP.

Expansion of lower-elevation ecosystems into higher elevations in Khuvsgul Lake National Park. Generally, in northern Mongolia, closed forests occur up to about 2,400 m elevation, and open forests occur up to about 2,900 m elevation (Heiner et al. 2017). The findings of the present study indicate that, in KLNP, forest ecosystems occur only where the maximum temperatures in July are higher than 14°C. As the climate in KLNP warms, the elevational distribution of areas with July maximum temperatures higher than 14°C will shift upward. This is likely to promote tree growth at higher elevations, and large-scale conversion of alpine vegetation to forest vegetation. The extent of such colonization by forests into alpine areas is likely to be affected by livestock grazing and timber cutting, which suppress forest growth.

Conversion of one type of ecosystem to another. The warming and drying climate will promote the growth of vegetation that favors warmer and drier conditions. The study findings indicate this may result in at least three projected scenarios, as follows:

- Conversion of closed forest ecosystems to open forest ecosystems. Impacts may include reduced tree density in open forests (i.e., open forests will become more "open") and expansion of open forest into alpine areas. Overall, projections include the decline of closed forest by about 45% (from 721 km^2 to 400 km^2, Table 14).
- Conversion of open forest to forest steppe, with trees persisting in locally favorable sites with larger areas of steppe vegetation in between.
- Conversion of the meadow ecosystem by 76% (from 145 km^2 to 35 km^2, Table 14) to shrub and steppe ecosystems, as soil and air moisture content declines.

For the four riverine ecosystems in KLNP (riverine forest, riverine meadow, riverine shrub, and small riparian stream), climate change will affect these ecosystems in different ways compared with the other terrestrial ecosystems. This is because the distribution, extent, and condition of riverine ecosystems are influenced particularly by the hydrology of their drainage basins, including the amount and timing of precipitation inputs. These were key factors in the regional mapping of these ecosystem categories (Heiner et al. 2017). The modeling in the current study indicates that these ecosystems may remain stable or may decline (Table 14), but did not include hydrological analysis at the level of the drainage basin for Khuvsgul Lake. For a better understanding of the impacts of climate change on these ecosystems, future modeling studies for KLNP should include hydrological analysis at the level of the lake's drainage basin.

In the long term, changes in the composition of flora and fauna communities due to climate change may eventually require adjustments in the existing definitions of ecosystems.

Table 14: Projected Changes in the Extent of Ecosystems and Bioclimatic Zones in Khuvsgul Lake National Park by 2050, Compared with the Baseline Conditions (1960–1990)

| Ecosystem (Area, km²) | Bioclimatic Zone (D, F, G) (% ecosystem) | | | Projected Change in Environmental Stratum by 2050[b] | Total Projected Change by 2050 |
	Primary Stratum[a]	Secondary Stratum[a]	Tertiary Stratum[a]		
Alpine barren (36)	F_06 (54)	F_03 (31)	D_03 (13)	Primary will decline to 17% (22 km²); secondary will become F_10 (unsuitable); tertiary will disappear.	Decline of 89%–92% (36 km² to 3-4 km²)
Alpine steppe (905)	F_08 (36)	F_03 (27)	F_05 (24)	Primary (38%, or 516 km²) will remain, the rest will become F_10 or G_01 (unsuitable); secondary and tertiary will mostly become F_10 (unsuitable).	Decline of 86% (905 km² to 130 km²)
Alpine tundra (169)	F_03 (76)	F_06 (14)		Primary will mostly become F_10 or F_08 (unsuitable); secondary will become F_08 (unsuitable).	Decline of 88% (169 km² to 20 km²)
Forest closed (721)	F_10 (57)	G_01 (40)		Primary will become G_01 (suitable); secondary will become G_06 (unsuitable).	Decline of 45% (721 km² to 400 km²); change to open forest
Forest open (1,974)	F_10 (65)	G_01 (25)		Primary will become G_01 (suitable); secondary will become G_06 (suitable).	Expansion into closed forest and alpine areas
Forest steppe (1,510)	F_10 (55)	G_01 (40)		Primary will become G_01 (suitable); secondary will become G_06 (suitable).	Expansion into closed forest and alpine areas
Lower mountain steppe (3,285)	F_10 (58)	F_08 (26)	G_01 (13)	Primary will become G_01 (suitable); secondary (38%, or 516 km²) will remain, the rest will become F_10 or G_01 (suitable); tertiary will become G_06 (appears unsuitable).	Loss of 310 km² at lower elevations; may be offset by expansion into alpine steppe
Meadow (145)	G_01 (75)	F_10 (25)		Primary will become G_06 (unsuitable); secondary will become G_01 (suitable).	Decline of 76% (145 km² to 35 km²)
Riverine forest (34)[c]	G_01 (83)	F_10 (17)		Primary will become G_06 (less suitable); secondary will become G_01 (suitable).	May persist or become riverine shrub
Riverine meadow (39)[c]	G_01 (94)			Primary will become G_06 (unsuitable).	Threatened by livestock grazing
Riverine shrub (68)[c]	G_01 (79)	F_10 (21)		Primary will become G_06 (suitable); secondary will become G_01 (suitable).	Stable or may expand
Small riparian stream (98)[c]	F_10 (50)	G_01 (40)		Primary will become G_01 (suitable); secondary will become G_06 (suitable).	Stable
Steppe (42)	G_01 (77)	F_10 (22)		Primary will become G_06 (suitable); secondary will become G_01 (suitable).	May expand into F_10 and new G_06

D = extremely cold and wet zone, F = extremely cold and mesic zone, G = cold and mesic zone, km² = square kilometer.

Notes:
1. Ecosystem categories are derived from Heiner et al. (2017).
2. Climate codes are derived from Metzger et al. (2013).
3. The Appendix provides details about the global model, zones, and strata relevant to this study.

[a] Primary stratum = stratum occupied by the largest percentage of the ecosystem in Khuvsgul Lake National Park; secondary stratum = stratum occupied by the second-largest percentage of the ecosystem that is >10%; tertiary stratum = stratum occupied by the third-largest percentage of the ecosystem that is >10%. See Chapter 4 for a description of the bioclimatic zones and strata.

[b] Suitable = the ecosystem currently occurs in this stratum; unsuitable = the ecosystem is currently not found in this stratum either in Khuvsgul Lake National Park or in surrounding areas.

[c] See text on modeling results for the riverine ecosystems (p. 29 of this report).

Source: Asian Development Bank.

Types of ecosystems that occur under warmer and drier conditions outside Khuvsgul Lake National Park. The entry of a new environmental stratum (G_06 of the cold and mesic zone) will bring warmer and drier conditions that will eventually occupy 42% (4,923 km²) of KLNP (Table 12). This will almost certainly cause large changes in vegetation communities and, therefore, in the integrity of ecosystems in KLNP. To further understand the potential nature of these changes, the ecosystems that currently occur under the bioclimatic conditions of stratum G_06 near KLNP were examined. Lands southeast of KLNP are at lower elevations and have warmer temperatures than KLNP and occur under the G_06 stratum. The projected climate parameters for this stratum include a mean maximum temperature of 2.7°C and an AWI of 0.51 by 2050. The ecosystems in this region comprise forest steppe (54%), open forest (22%), and riverine shrub (13%). There is no closed forest and almost no riverine meadow or small river riparian ecosystems. Further south in Mongolia, the steppe ecosystem is common in this stratum. The existing stratum (G_01) that will be replaced by the G_06 stratum (Table 12) is characterized by a lower mean temperature of 1.5°C and an AWI of 0.50. The occupation of much of KLNP by the G_06 stratum is likely to cause the conversion of closed forest to open forest or forest steppe, the conversion of riverine meadow to steppe ecosystems, and a decline in the extent of riparian ecosystems.

These projections do not account for local differences in climatic and environmental conditions, or the linked effects of human activities, especially livestock grazing, timber cutting, and fire (Addressing Existing Threats, Chapter 6). Satellite imagery of KLNP and ground truthing conducted for the study indicate that, in river valleys where livestock grazing is excluded, vegetation conditions are notably different from those in areas where grazing occurs. In such places, floodplains exhibit lush, dense steppe vegetation with a thick cover of trees and shrubs on lower slopes. There may be significant movement of forests into valleys in those areas. Detailed mapping of permafrost and vegetation communities would enable better modeling of spatial changes in the moisture regime and how these changes may influence tree distribution.

Compounded impacts of existing threats and climate change on biodiversity. Except for those species of plants and animals restricted to the highest, most remote alpine habitats (for which climate change may be the greatest threat), most species in KLNP occur at lower elevations and are already being impacted by human activities, e.g., hunting, fishing, and livestock grazing. The need to address these existing impacts is a high priority to build resilience to climate change. For example, KLNP is a national stronghold for red deer (*Cervus elaphus*), a species that is critically endangered in Mongolia largely because of hunting (Clark et al. 2006). The species is mobile and uses a range of habitats (Mattioli 2011). Mapping of the regional distribution of red deer (Heiner et al. 2017) indicates that the species occurs in at least five ecosystems in KLNP (closed forest, open forest, lower mountain steppe, forest steppe, and alpine steppe), of which at least three may undergo large declines due to climate change (Table 14). In the short term, the highest management priority for red deer at KLNP may be to address hunting and, in the longer term, to improve the protection of forest ecosystems from livestock grazing. Together, these measures may be the most important approaches to strengthening the resilience of local red deer populations to changes in feeding and/or breeding habitats that will occur under climate change. Similar strategies are likely to apply for other large, threatened grazing mammals in KLNP.

Khuvsgul Lake

Because of its large size and distinctiveness as a key landscape unit within KLNP, Khuvsgul Lake is an ecosystem in itself and warrants assessment for the unique aquatic conditions it supports compared with the surrounding terrestrial ecosystems.

Under the modeled baseline (1960–1990) climate, the lake was located in a single environmental stratum (G_01) of the cold and mesic zone. By 2050, the lake is projected to be under an entirely new environmental stratum (G_06) that does not occur at present within KLNP (Figures 10 and 11). This change in environmental strata is projected to result in an increase in air temperatures (with minimum temperatures 0.6°C higher and maximum temperatures 1.6°C higher), and increased water loss by evaporation, as annual precipitation could decrease by about 68 mm and PET could increase by about 57 mm. The number of GDD is projected to increase by 287, from 1,285 (Table 2) to 1,572. These findings were applied, together with information from other studies, to briefly assess potential impacts on five dimensions of the lake's hydrology and ecology: water quality, hydrology, thermal stratification, physical condition, and aquatic biodiversity. Modeling of the impacts of climate change on the hydrology of the lake watershed was beyond the scope of the study, yet such impacts are likely to have direct effects on the lake and should be considered in future studies of Khuvsgul Lake.

Water quality. Khuvsgul Lake is characterized by clear water with low nutrient levels and a catchment geology with high levels of phosphorus (Urabe et al. 2006). Water temperatures are naturally low and rates of bacterial decomposition are slow. This contributes to the lake being especially susceptible to pollution. Water quality is already threatened by high nutrient loads from livestock waste entering the lake and sewage outflow from nearby tour camps, which have already contributed to algal blooms and fish deaths (Goulden and McIntosh 2018). Climate change is causing the lake waters to become warmer (Chapter 3), and this presents a major additional threat to the lake's water quality and ecological values.

The impacts of increasing nutrient loads and rising lake temperatures on lake water quality and ecology will be most severe, and occur the soonest, in small bays around the main lake. Numerous small, semi-enclosed bays with narrow outlets occur around the lake and can be considered almost as separate, smaller lakes. Although their surface water layers can mix with the main lake, a sill or ridge at the outlet reduces mixing of the lower water layers. This results in a stratified bottom layer of water that can become anaerobic, causing high levels of sedimented phosphorus to dissolve and accumulate in the lower water layer, which eventually mixes with the upper water (C. E. Goulden, in litt.). This mixing stimulates algal blooms and fish deaths, and is most common in bays along the lake's western shore, from mid to late summer (C. E. Goulden, in litt.). Warmer temperatures result in the increased release of phosphorus trapped in the lake sediments (Goulden and McIntosh 2018, and references therein). As lake temperatures rise and nutrient loads from livestock and sewage increase, this cycle of impacts will increase. In the short term, these impacts may not affect the main lake, but are of particular concern for bays that support tour camps.

Hydrology. Between 1992 and 2008, the area of Khuvsgul Lake expanded by about 54.9 km² (Oyuungerel and Munkhdulam 2011), equivalent to about 2% of the size of the lake. This expansion is attributed to permafrost melt (around the lake) and melting of the park's single glacier (situated north of the lake), due to climate change and livestock damage to vegetation and soil (Oyuungerel and Munkhdulam 2011). Other glaciers in the region are already experiencing high levels of melting (Zhang et al. 2017), and the single glacier at KLNP is likely to disappear in the near future. In the longer term, the projected increases in air and water temperatures, reduced precipitation, and increased rates of evapotranspiration at KLNP (Chapter 3) may eventually result in reduced water inputs to Khuvsgul Lake and changes in the seasonal timing of water inputs to the lake (with more inputs earlier in summer, followed by a longer period of dry conditions). Reduced water inputs to the lake will also result in reduced outflows from the lake. This would affect water supply across the lake's watershed, which is part of the Selenge River system and extends across northern Mongolia before draining into Baikal Lake in the Russian Federation.

Thermal stratification. The water body of Khuvsgul Lake exhibits distinct layers of warmer and colder water, which overturn twice a year (termed "dimictic"). Both occasions occur in summer, in May–June (in shallower waters near the shores) and July (in deeper water) (Hayami et al. 2006). This seasonal reversal of the water temperature profile is critical to the distribution of nutrients between shallow and deep areas and the provision of food for fish and invertebrates. The projected increases in water and air temperatures and earlier spring thaw of the lake ice (Chapter 3) could result in the water turnover occurring earlier in summer. This would create warmer and more nutrient-rich conditions, which would last for a longer duration in summer. This would, consequently, increase the risk of algal blooms—especially when compounded by the addition of nutrient loads from tourism, which is highest in summer—and fundamental alterations in the aquatic ecology of the lake.

Physical condition. Storms are increasing in frequency at KLNP (Chapter 3). Wave action generated by wind is a natural feature of Khuvsgul Lake, but more frequent and/or larger storms may generate larger waves that cause greater physical damage to lake shores and vegetation. This would be compounded by increased erosion due to larger storm events over a longer ice-free period, as well as vegetation loss from overgrazing and human disturbance. Rising temperatures, combined with a change from more frequent, light rains (which soak the earth and help maintain soil moisture) to less frequent but heavier rainfall (which causes higher runoff and damage), would contribute to more rapid drying of soils and wind-blown erosion. In turn, this could affect the growth of vegetation and cause greater soil loads to enter the lake.

Aquatic biodiversity. Khuvsgul Lake supports communities of unique aquatic invertebrates (Safronov 2006; Sitnikova, Goulden, and Robinson 2006) and 10 species of fish (Free, Jensen, and Mendsaikhan 2016). One fish species, the Khuvsgul grayling, is unique to the lake and the population of another (the lenok, *Brachymystax lenok*) is biologically distinct, rendering the lake of global importance for this population (Kaus et al. 2019). The KLNP also supports populations of at least two amphibian species ranked as vulnerable in Mongolia, the Siberian salamander (*Salamandrella keyserlingii*) and the Japanese tree frog (*Hyla japonica*) (Terbish et al. 2007), which may occur in shallow wetlands and streams around Khuvsgul Lake.

Little is known about the ecology of most plant and animal species in Khuvsgul Lake, but the lake's isolation and the uniqueness of many species clearly indicate that these communities have evolved to suit the lake's conditions— i.e., clear, clean, and cold waters with complex cycles of seasonal warming and cooling and nutrient distribution. Climate change may fundamentally alter these cycles and cause large changes in the lake's ecology. As climate change may occur across the entire lake, the ecological impacts that occur are likely to be large. The lake is also vulnerable to climate-induced impacts on the surrounding terrestrial ecosystems, including increased soil loads resulting from increased upstream erosion and melting of permafrost.

For aquatic invertebrates, declining water quality (due to lake pollution from livestock grazing and tourism, combined with increased algal blooms in the warming conditions) and changes in the seasonal timing of the overturning of the lake waters present severe threats. For other cold mountain lakes, these issues have caused changes in the composition of invertebrate communities (Shimaraev and Domysheva 2013) and increased numbers of smaller fish (Jeppesen et al. 2013). At Khuvsgul Lake, such changes may result in the extinction of species and also affect food supply for other species. For the fish in Khuvsgul Lake, climate change is identified as a key threat (Olson et al. 2019; Tsogtsaikhan et al. 2017). The ecology and daily and seasonal movements of fish in the lake are linked with the distribution of their invertebrate prey (Ahrenstorff et al. 2012), and some species spawn upstream (Sideleva 2006). Changes in the thermal regime of the lake, declines or changes in prey availability, and reduced stream flows could reduce food availability and disrupt the ecological cycles for breeding and foraging. These impacts will compound the effect of overfishing, which is a key existing threat that is already causing a decline in fish populations in Khuvsgul Lake (Ahrenstorff et al. 2012; Free, Jensen, and Mendsaikhan 2016).

These issues indicate that extensive adverse impacts are likely for the aquatic biodiversity of Khuvsgul Lake. Combined with lake pollution, this may result in the extinction of species that can no longer survive or compete under the changed conditions. Because of the isolated nature of the lake, its plants and animals have few options for adaptation, e.g., shifting northward to cooler waters (Olson et al. 2019; Tsogtsaikhan et al. 2017). This emphasizes the importance of management actions that address the existing threats to the lake (Chapter 6).

Livelihoods

The assessment of climate change impacts on livelihoods focused on livestock herding, the dominant livelihood activity in KLNP (Chapter 2). The distribution of herder camps was mapped to help assess the potential impacts of climate change on herding livelihoods as well as the impacts of grazing on vegetation communities. In 2019, about 821 herder families used 1,023 camps in and close to KLNP; 988 of these camps were mapped.

A total of 688 of these camps were within KLNP. Most of the camps in KLNP (650 of mapped camps, or 66%) were located within 10 km of Khuvsgul Lake, along the northeast, northwest, southeast, and southwest shores (Figure 1). These findings are consistent with previous studies, which indicate that virtually all shorelines and lowlands and low valleys around Khuvsgul Lake are used for livestock grazing. The location of herder camps in KLNP in 2019 was overlaid with the bioclimatic zones and ecosystems under the baseline climate (1960–1990). The key findings were as follows:

- In 2019, 477 of 688 camps (69%) were located in the cold and mesic zone (stratum G_01) and 211 camps (31%) were in upper valleys of the extremely cold and mesic zone (stratum F_10). The other 300 camps outside KLNP were located in the strata G_01 (225; 75%), G_06 (24; 8%), and F_10 (50; 17%).
- The dominant ecosystems in these strata are lower mountain steppe and forest steppe, with smaller areas of riverine shrub and small riparian streams. Because of the projected loss of stratum G_01 and its replacement with G_06 (Bioclimatic Zones, this chapter), riverine ecosystems may decline and steppes may replace lower mountain steppes.
- The reliance on lower valleys for livestock grazing (stratum G_01) indicates that these will be the first and most heavily impacted locations under climate change.
- There is a high level of awareness among residents about the threats posed by climate change to herding. Of 700 households interviewed, 630 (90%) stated that climate change was contributing to pasture degradation, and 574 (82%) said that this was affecting income and causing a decline in livestock productivity. These results are similar to those from a survey of 96 herders in 2009–2010, who reported that rains had changed (98%) and were more intense (89%), colder (82%), and/or shorter (61%), and that heavier rainfall was damaging pasture and increasing hazards to families while tending herds (Goulden et al. 2016).

The impacts of climate change on herding livelihoods are closely linked with livestock grazing and fire, and their collective impacts on pasture, soil moisture content, and melting of permafrost. Unsustainably high rates of livestock grazing have already resulted in widespread damage to vegetation and pasture (Livelihoods, Chapter 6). Climate change will further threaten livelihoods by compounding these impacts. Higher rates of evapotranspiration, and the concentration of rainfall into more intense but less frequent events, will contribute to declining pasture conditions and the drying of springs and wetlands. The projected increases in the area of open forest, forest steppe, and lower mountain steppe ecosystems (Table 14) due to warming conditions could benefit livelihoods by increasing the area of pasture for livestock; but, without improved livestock management, such benefit is unlikely.

The impacts of climate change on livelihoods are likely to be most severe for poor households, which have a low inherent resilience to change because of limited opportunities for income diversification. Reduced livestock productivity resulting from climate change may result in more hunting and fishing (to supplement food and income), placing further pressures on biodiversity.

Tourism

To help assess the potential impacts of climate change on tourism in KLNP, the location of 72 mapped tour camps in KLNP in 2019 was overlaid with the bioclimatic zones and ecosystems under the baseline climate (1960–1990). The key findings are as follows: (i) all tour camps are located in one environmental stratum (G_01), and the dominant ecosystems in this stratum are forest steppe (28% of the non-lake area of the stratum), open forest (26%), lower mountain steppe (19%), closed forest (13%), meadow (5%), and other non-alpine ecosystems (small percentages); (ii) tour camps are located mainly in six types of treeless ecosystems—lower mountain steppe (30 camps, or 42% of the total), forest steppe (20; 28%), meadow and riverine meadow (15; 21%), steppe (5; 7%), and open forest (2; 3%); and (iii) 60 camps (83%) are within 500 m of Khuvsgul Lake (Figure 1).

Tourism at KLNP has been largely unmanaged and this is affecting the park's ecological and scenic values (Tourism, Chapter 6). Climate change will compound these impacts. In addition, the overlay of tour camp locations on the modeled bioclimatic zones under baseline conditions indicates a climate-related threat to two ecosystem categories—meadow and steppe. In 2019, a total of 35 tour camps (49% of the total) were located in these ecosystems. Both ecosystems occur in only small proportions in the environmental stratum G_01; under climate change, the extent of at least one of these ecosystems (meadow) will significantly decline (Table 14). Improved planning is required for tour camp locations to avoid further development in this ecosystem and maximize its resilience to climate change.

The potential impacts of climate change on tourism were assessed through three perspectives: (i) damage to tourism infrastructure; (ii) hazards to visitors and residents; and (iii) declining quality of the visitor experience, combined with reduced opportunities for local income from tourism.

Damage to tourism infrastructure. Infrastructure in KLNP includes a small road network, tour camps, residential homes, powerlines, a visitor center, park headquarters, and small jetties and bridges. Climate change may cause increased damage to structures from melting permafrost and more extreme weather events, and may require higher investments for the design, operation, and maintenance of infrastructure. Melting permafrost due to warming conditions has already caused the collapse of buildings in Khatgal town (Sharkuu 2006). Because of the wide distribution of permafrost in KLNP (Chapter 2), many structures are vulnerable to land subsidence from melting permafrost. Other risks include damage to jetties and piers due to stronger wind and wave action, the weakening of structural foundations due to bank erosion under heavier rains, and fire damage under drier conditions. In a survey of 51 tour operators (Appendix), thunderstorms were reported to have caused flooding at 17 tour camps, and melting permafrost was reported to have caused structural damage and tree loss at 11 camps.

Hazards to visitors and residents. Increased hazards to the safety of visitors and residents may result from extreme weather (e.g., flash events of storms, floods, heat waves) and increased risk of disease outbreaks under warmer conditions. Tour camps are especially vulnerable to climate hazards as most are located close to the shore of Khuvsgul Lake and nearby forest stands, and are exposed to the risks of storms, wind, wave action, and fire. Warming conditions could also increase the risk of the coronavirus disease (COVID-19) and/or other disease outbreaks among residents and visitors in KLNP, especially when combined with low sanitation standards. As of July 2020, no cases of COVID-19 had been detected in Khuvsgul *aimag*. Mongolia had among the lowest COVID-19 infection rates in the world (WHO 2020), largely because of the rapid closure of international borders. As tourism resumes, the risk of disease outbreaks at KLNP will increase. The impact of such outbreaks at KLNP could be severe, as there are few medical facilities in the region.

Quality of the visitor experience and income from tourism. Climate change may impact the quality of visitor experience at KLNP and contribute to declines in visitor numbers. Negative visitor experiences may result from the pollution of Khuvsgul Lake (due to poor water quality, compounded by rising waters and algal blooms) and climate hazards. Khuvsgul Lake is the centerpiece of tourism at KLNP, and degraded lake conditions that no longer display pristine, clear waters would likely result in disappointment and complaints. Low sanitation standards are already rated by visitors as a factor affecting the tourism experience in Mongolia (Yu and Goulden 2006), and the increased risk of disease outbreaks under warming conditions could further affect visitor perceptions and concerns about travel to KLNP. In other countries, lake pollution due to tourism and unregulated development has resulted in health impacts, the closure of lakes for recreational use, and the need for extensive cleanups and financing, including the construction of wastewater treatment infrastructure (e.g., Cooke 2007). Under climate change, these risks will be highest for the bays around Khuvsgul Lake, which are occupied by tour camps, and in summer, when algal blooms are most common and visitor numbers are highest.

Lower visitor numbers would reduce the opportunities for residents to derive income from tourism. This is significant as nature-based tourism is one of the few livelihood opportunities compatible with conservation in KLNP. Climate change may also affect the opportunity to expand tourism beyond the peak summer season. An "ice festival" has been held annually at KLNP since 2000 and is one of the few winter tourism events in Mongolia. The event is held in early March, on the frozen surface of Khuvsgul Lake, and relies on a stable ice layer and clear, sunny conditions. Climate change is causing the frozen lake surface to begin thawing before March (Chapter 3); combined with more extreme weather, this could affect the festival timing as well as visitor safety.

Park Management

The potential impacts of climate change on the management of KLNP were assessed through (i) modeled changes in the proportion of ecosystems within the three management zones of KLNP by 2050, compared with baseline conditions (Table 15); and (ii) consideration of potential climate impacts on the objectives of the KLNP management plan for conservation.

Khuvsgul Lake National Park management zones. Under the modeled baseline climate (1960–1990), three management zones of KLNP—special zone, limited use zone, and tourism zone—were studied.

- *Special zone.* Almost all alpine ecosystems are located in the special zone (they receive the highest level of administrative protection). This location is significant, as these ecosystems are the most threatened by climate change. Five other ecosystems (closed forest, open forest, lower mountain steppe, riverine forest, and small riparian stream) are relatively well protected, with about 70% or more of each ecosystem located in the special zone.
- *Limited use zone.* A large proportion of the ecosystems found at lower elevations are in the limited use zone.
- *Tourism zone.* All of Khuvsgul Lake is zoned for tourism. This is of conservation concern for at least four ecosystems (Khuvsgul Lake, meadow, riverine meadow, and riverine shrub). For Khuvsgul Lake, there is a risk of overdevelopment and reduced opportunities to build resilience to climate change by protecting shoreline vegetation. For the other three ecosystems, they already have a restricted extent in KLNP and are threatened by grazing and human activity.

Table 15: Proportion of Ecosystems within the Management Zones of Khuvsgul Lake National Park, under the Baseline Conditions (1960–1990)

Ecosystem Category	Area in KLNP (km²)	Proportion of KLNP Area (%)	Representation of Ecosystem within Management Zone (%)			Proportion of Management Zone Represented by Ecosystem (%)		
			Special Zone	Limited Use Zone	Tourism Zone	Special Zone	Limited Use Zone	Tourism Zone
Alpine barren	36	0.3	100.0	0.0	0.0	0.5	0.0	0.0
Alpine steppe	905	7.7	99.6	0.3	0.1	13.2	0.2	0.0
Alpine tundra	169	1.4	99.5	0.5	0.0	2.5	0.0	0.0
Dry river	2	0.0	0.0	53.5	0.0	0.0	0.0	0.0
Forest closed	721	6.1	90.0	7.8	3.0	9.5	3.0	0.7
Forest open	1,974	16.7	73.7	22.8	3.4	21.4	24.1	2.2
Forest steppe	1,510	12.8	55.8	35.4	8.2	12.4	28.6	4.0
Lower mountain steppe	3,285	27.8	79.2	18.1	1.8	38.2	31.8	1.9
Meadow	145	1.2	12.9	70.6	1.4	0.3	5.5	0.1
Riverine forest	34	0.3	69.9	25.9	7.3	0.3	0.5	0.1
Riverine meadow	39	0.3	13.3	61.4	1.1	0.1	1.3	0.0
Riverine shrub	68	0.6	32.2	59.9	3.3	0.3	2.2	0.1
Small riparian stream	98	0.8	68.7	26.1	3.2	1.0	1.4	0.1
Steppe	42	0.4	27.0	54.9	1.1	0.2	1.2	0.0
Khuvsgul Lake	2,781	23.5	0.0	0.0	100.0	0.1	0.3	90.0
Total	**11,808**	**100.0**				**100.0**	**100.0**	**100.0**

KLNP = Khuvsgul Lake National Park, km² = square kilometer.
Note: Numbers may not sum precisely or percentages may not total 100% because of rounding.
Source: Asian Development Bank.

The extent to which the proportion of each ecosystem within the three management zones may change by 2050 is unclear, even assuming that no revisions in the zoning are made. However, the large projected changes in the types and extent of environmental strata in KLNP (Tables 12 and 14) indicate that large changes in the representation of each ecosystem within each zone are likely. This emphasizes the need to revise the park zoning to maximize the administrative protection of threatened ecosystems.

Climate change and conservation targets. The Khuvsgul Lake National Park Management Plan, 2015–2020 (MEGDT 2014b) and the Integrated Water Resource Management Plan for the Khuvsgul Lake and Eg River Basin (MEGDT 2014a) identify the need to assess climate change impacts and develop adaptation measures for KLNP. The KLNP management plan identifies nine targets to represent the conservation values of KLNP: Khuvsgul Lake, mineral springs, one plant species (*S. involucrata*), two vegetation communities (mountain steppe plant communities and larch forest), one fish species (Lenok), one bird (white-tailed eagle, or *Haliaeetus albicilla*), and two mammal species (moose, Siberian ibex). The plan ranks (as *high*, *average*, or *low*) the relative importance of 12 threats to the targets, including climate change. Climate change is listed as an *average* threat to four targets (Khuvsgul Lake, mineral springs, *S. involucrata*, and moose) and a *low* threat to one target (mountain steppe plant communities). Illegal fishing, overharvesting, livestock grazing, and hunting are ranked as *high* threats to these various targets. No climate ranking is accorded to the other targets.

The study findings only partly support these climate rankings. For three values (Khuvsgul Lake, mineral springs, and *S. involucrata*), the relative threat of climate change should probably be elevated. This reflects the projected shift to a warmer and drier climate across KLNP, which may especially threaten small streams and shallow areas of Khuvsgul Lake, and the severe projected decline in the alpine habitat of *S. involucrata* due to climate change (subsections Terrestrial Ecosystems and Khuvsgul Lake under Ecological Values, this chapter). For one target (Lenok), no climate ranking is given, but the threat assessment (Khuvsgul Lake, under Ecological Values, this chapter) indicates that a ranking of at least *low* for climate change is warranted. For mountain steppe plant communities (*low* climate threat) and larch forest (no climate ranking), the study findings accord with the KLNP rankings, as the projected changes in the forest steppe and lower mountain steppe ecosystems (Table 14) indicate lower climate impacts. For the remaining targets, there is insufficient information to assess the validity of the KLNP rankings; although, for moose, Siberian ibex, and white-tailed eagle, habitat loss and disturbance may be greater immediate threats than climate change.

Management effectiveness. The study findings indicate that climate change will have an impact on the effectiveness of management interventions in KLNP. Measures to address climate change must therefore be incorporated into park management, together with measures to address existing threats to conservation values. These issues are discussed in Chapter 6.

6 Building Resilience to Climate Change

Key Messages

- For protected areas, three approaches are generally recognized as necessary to build resilience to climate change: address existing threats, improve habitat connectivity, and strengthen management. These approaches apply to Khuvsgul Lake National Park (KLNP).
- For KLNP, the highest management priority to build resilience to climate change is to address two key issues: excessive livestock numbers and unmanaged tourism. These issues have caused widespread damage to habitats and the pollution of Khuvsgul Lake. Climate change will further compound these impacts.
- Measures to address these issues and other identified threats are recommended. These include measures for biodiversity conservation and the management of livestock grazing and pasture, tourism, sanitation, and waste control.
- The KLNP is located in a landscape that is well suited to transboundary conservation. Nearby regions of northern Mongolia and the Russian Federation support a mosaic of protected areas and limited development. With effective planning, these attributes can enable plants and animals in KLNP to shift to new habitats as climate conditions in the park become unsuitable for them.
- These issues are complex and present major challenges for the management of KLNP. Addressing these issues will require institutional reform, the revision of management priorities and approaches, and technical and financial support.

Building resilience to climate change comprises the design and implementation of actions that reduce the vulnerability of biological and social systems to change. These actions are termed adaptation practices (see Glossary). For protected areas, three approaches are generally recognized as necessary to building climate resilience: addressing existing threats, maintaining and enhancing habitat connectivity, and improving management effectiveness (Hole et al. 2009). These approaches are applied to Khuvsgul Lake National Park (KLNP) as follows.

Addressing Existing Threats

Ecological Values

Reviews of management priorities for KLNP generally highlight seven existing local threats to biodiversity: livestock overgrazing, unmanaged tourism, overfishing, hunting, timber extraction, fire, and climate change (IUCN 2008; Goulden and McIntosh 2018). High livestock numbers are causing severe damage to vegetation, changes in the composition of vegetation communities, pollution of Khuvsgul Lake, and physical damage to sensitive stream and shore habitats. Most ecosystems in KLNP are impacted by overgrazing (Ecosystems, Chapter 4). Combined with rising temperatures and drier conditions, this is contributing to a continuous cycle of tree dieback, drier soils, melting permafrost, fire, and conversion of forest to steppe (Goulden et al. 2006). Rapid and largely unmanaged tourism in KLNP has emerged as a major threat to biodiversity (Tourism, this chapter). Fires, which are largely caused by human carelessness (Nyamjav, Goldammer, and Uibrig 2007), and timber extraction compound grazing impacts by limiting the regeneration of trees and shrubs. Fish populations in Khuvsgul Lake are declining because of overfishing (Khuvsgul Lake, under Ecological Values, Chapter 5). Hunting (especially of large mammals), collection of non-timber forest products, and timber extraction are increasing (IUCN 2008).

Previous studies, and the findings of the present study (Chapter 5), indicate the following:

- At the large scale of ecosystems and landscapes, the principal threats to biodiversity in KLNP are unsustainable levels of livestock grazing and unmanaged tourism. These issues are driving declines in the extent and quality of vegetation and habitats across much of the park.
- At the finer level of individual species, for most fish species, large mammals, and some plants, overextraction—from overfishing, hunting, or collection—appears to be the highest immediate threat to local populations.
- Climate change will compound these impacts by causing large projected changes in the types and extent of ecosystems in KLNP (Ecological Values, Chapter 5).

Based on these findings, the highest priority to build climate resilience for most plant and animal species in KLNP is to focus on addressing these existing threats.

The KLNP management plan lists the need to control illegal fishing and hunting and prioritizes the protection of selected flora and fauna, although the basis for the selection of some of these species is unclear (Park Management, Chapter 5). Existing management measures by the KLNP Administration include conducting ranger patrols; monitoring the populations of some species; and working with communities to reduce fishing, hunting, and the risk of fires. Additional measures to address existing threats to biodiversity in KLNP should include the following:

- Reduce overgrazing pressure and improve livestock and pasture management through measures that benefit biodiversity and livelihoods (Livelihoods, this chapter).
- Improve the planning and management of tourism, especially to reduce pollution of Khuvsgul Lake (Tourism, this chapter).
- Prioritize the allocation of limited management resources and conduct an updated and systematic assessment of the threats to flora and fauna of KLNP. This assessment should include, at a minimum, all rare, threatened, and restricted-range plant and animal species in KLNP. It should be based on scientific data, expert input, and community knowledge, and include the findings of this study on climate change.
- For fish, (i) expand the list of priority species in the KLNP management plan to include the Khuvsgul grayling (the only endemic fish species in Khuvsgul Lake); and (ii) conduct an updated assessment of fishing activities and develop measures to protect and manage fish communities, including sustainable catch levels for selected species.
- For high-altitude plant species at risk from climate change, e.g., *S. involucrata* (Ecological Values, Chapter 5), halt existing harvesting and exclude livestock grazing from sites known to support this species.
- Strengthen the enforcement of park regulations for hunting, fishing, and the collection of timber and non-timber forest products. This should include improved collaboration between the KLNP Administration and communities, and the strengthening of the management capacity of park staff (Strengthening Park Management, this chapter).

These measures will support the recovery of plant and animal populations and their habitats in KLNP and, in doing so, strengthen the resilience of local populations to climate change.

Livelihoods

Existing threats to herding livelihoods in KLNP comprise unsustainably high livestock numbers and few options for alternative incomes. In 2019, over 99,000 livestock were present in KLNP. These consisted of goats (about 34,170), sheep (32,166), cattle (25,166), and horses (8,159) (KLNP Administration, unpublished data). Herders traditionally grazed their livestock around Khuvsgul Lake in the summer, then moved to upper mountain valleys 10–20 km away for winter grazing. Because of increasing herd sizes, competition for grazing lands, and a scarcity of high-quality pasture, most herders now permanently graze their livestock around Khuvsgul Lake (Goulden et al. 2016; Livelihoods, Chapter 5). This practice has led to overgrazing, soil erosion, and declining pasture productivity (Enkhtaivan 2014). These changes have impacted livestock health, increased the vulnerability of livestock to harsh weather conditions, and reduced the ability of herders to prepare adequate winter feed reserves. In 2009–2010, these factors, combined with a harsh winter, contributed to the mortality of about 28,000 livestock (KLNP Administration, unpublished data).

Guidelines for Mongolia's livestock sector (Batima 2006) to address these issues as well as to increase resilience to climate change include (i) increasing seasonal reserves of winter fodder, through setting aside more pasture lands as reserves and increasing hay making and nonplant feed preparation; (ii) expanding credit access for herders to enable them to purchase high-quality livestock, feed, and equipment for pasture improvement; (iii) providing training in livestock and pasture management to build on local knowledge and raise awareness of environmental degradation and climate change; and (iv) improving forecasting and warning systems for drought and harsh winters. In KLNP, herders have identified the need to collect more winter fodder, move livestock more frequently between grazing sites, and reduce herd sizes (Goulden et al. 2016). During the present study, government agencies and civil society groups also identified the need to address overgrazing and improve the resilience of herders to climate change by reducing herd sizes, improving herd health, and providing technical support.

Livestock overgrazing and the need to improve pasture management are identified in the KLNP management plan as priorities for management (MEGDT 2014b). As a protected area, measures to support herding livelihoods in KLNP need to be aligned with management objectives for biodiversity conservation. This presents additional challenges compared with the planning of livestock and pasture management outside protected areas. For example, goats are a preferred livestock in Mongolia because of income generation from cashmere, yet they cause severe grazing impacts compared with other livestock; their presence in KLNP is, therefore, not compatible with conservation objectives (Goulden and Goulden 2013).

Measures to address existing threats to herding livelihoods in KLNP, while supporting conservation objectives, should include the following:

- Establish herding groups to improve coordination and land planning between the KLNP Administration, herders, and other stakeholders.
- Establish pasture management plans for the herder groups to clearly delineate grazing areas and seasonal grazing regimes, and improve pasture health.
- Support herders in shifting livestock production from a focus on high livestock numbers to fewer, healthier livestock.
- Provide community training and financial support to (i) improve pasture management; (ii) improve the quality of meat and dairy products from livestock (to yield higher returns from fewer livestock); and (iii) develop alternative livelihoods, compatible with the park conservation objectives, to reduce dependence on herding.
- Implement the phased removal of goats from KLNP. This process should be planned with herders and accompanied with measures for livelihood support.
- Restrict livestock access to selected water points to reduce damage to streams and the pollution of Khuvsgul Lake from livestock waste.
- Exclude grazing from alpine ecosystems and other threatened ecosystems in the park, especially those most threatened by climate change (Ecological Values, Chapter 5).
- Improve collaboration between the KLNP Administration and herders in developing and implementing these measures.

These measures will help improve livestock and pasture health, the availability of winter fodder, and herder resilience to drought and harsh winters. This will support the recovery of overgrazed plant communities and fauna habitats, reduce physical damage to streams and pollution of Khuvsgul Lake from livestock waste, and improve the institutional framework for conservation and livelihoods in KLNP. The regeneration of vegetation will help strengthen the protection of soil moisture and permafrost, reduce the risks of soil erosion and fire, and increase the resilience of vegetation to drought as well as damage from larger storms.

Tourism

Managed sustainably, tourism can increase community resilience to climate change through income diversification and provide new revenue streams for park management. In contrast, tourism at KLNP has occurred with limited planning and is now a major threat to ecological values. Tourism is causing the pollution of Khuvsgul Lake due to the leakage of effluent from unlined pit toilets, dust raised by vehicles (which causes the dispersal of nutrient-rich soil), and inadequate waste facilities. These issues have resulted in declines in water quality (Natural Sustainable 2019) and high litter volumes (Free et al. 2014) in Khuvsgul Lake. These inputs compound the pollution caused by livestock waste and cause disproportionately high impacts on water quality because of the lake's sensitivity to nutrient inputs (Khuvsgul Lake, under Ecological Values, Chapter 5). Unregulated visitor access and infrastructure construction (e.g., tour camps and power lines) have caused damage to the lakeshore. In 2020, 55 tour camps were located within 200 m of Khuvsgul Lake; of these, 18 were within 50 m (Enkhtaivan and Munguntulga 2019, 80–81). The close proximity of such camps to the lake does not comply with minimum legal distances (Goulden and McIntosh 2018). Overfishing is driven by visitor demand. Tourism has also yielded few benefits for communities, as most residents have limited resources to develop tourism goods or services or lack the skills to work in tour camps.

Future developments, especially for tourism and road access, also present major threats to KLNP. The road network in KLNP is limited at present, and most roads are in poor condition. The only means of access to Khankh town and the park's northern region is via an unsealed road from Khatgal town, or a highway from the Russian Federation (Figure 1). Until recently, these conditions have helped protect the park's interior by limiting tourism development and visitor activity, but this situation is changing. In 2016, the main road from Khankh *soum* to the international border with the Russian Federation (Figure 1) was upgraded. There are unconfirmed plans for at least two other major road upgrades: (i) from Khankh *soum* south to Khatgal town (Goulden and McIntosh 2018), and (ii) from the western shore of Khuvsgul Lake west into Renchinlkhumbe *soum* (McIntosh 2017). Both would have large and likely irreversible impacts on KLNP, by increasing human activity and development in the core areas of the park.

Climate change will compound the impacts of tourism on the ecological values of KLNP, especially Khuvsgul Lake, as well as impacts that arise from future development. It will also undermine efforts to manage tourism sustainably and may reduce opportunities for communities to derive benefits from tourism (Tourism, Chapter 5). The KLNP management plan (MEGDT 2014b) prescribes the following actions to manage tourism: (i) limit tour camp development to selected lands around Khuvsgul Lake; (ii) minimize the density of tour camps; (iii) establish public infrastructure (for power, wastewater treatment, transport, and visitor facilities); (iv) improve tour camp services and sanitation; and (v) support community-based goods and services.

Tourism planning for KLNP under climate change will also need to consider sector recovery and resilience to COVID-19 and other diseases. The direct impact of COVID-19 on human mortality in Mongolia has been relatively minor (Tourism, Chapter 5), but the abrupt halt in tourism in 2020 due to national and international travel restrictions is estimated to have caused a decline in national 2020 tourism revenue of $101 million to $208 million (Abiad et al. 2020). For KLNP, visitor numbers for at least the 2020—and probably 2021—summer seasons are likely to be low. This will impact the livelihoods of tour operators and residents. As the immediate global impacts of COVID-19 recede and Mongolia's tourism resumes, there will be a need for integrated sector planning for tourism, health, and protected area management.

Additional measures to improve the management of tourism should include the following:

- Establish a dedicated agency to manage tourism in KLNP. The agency's role should be to support the KLNP Administration by promoting, managing, and monitoring tourism in KLNP; comply with laws and regulations; prevent pollution; protect ecological values; benefit communities; and generate revenue streams to be used for park management, conservation, and livelihoods. The agency should comprise, at a minimum, qualified personnel in nature-based tourism, waste management, and business administration. Such an agency would provide urgently needed expertise for KLNP and help build climate resilience by improving the management of tourism.

- Prepare a tourism, sanitation, and waste management plan for KLNP. This should include time-based targets (e.g., by a specified year, all tour camps will have adopted minimum waste control measures and facilities); limits to tour camp numbers; permitted locations for tourism infrastructure; and management and monitoring actions. Spatial planning for tour camps should comply with legal minimum distances to Khuvsgul Lake and avoid at least one small ecosystem (meadow) that has already been damaged by tour camp development and is projected to decline further under climate change (Table 14, Chapter 4). The plan should be integrated with recommended plans for waste control and road management (Strengthening Park Management, this chapter).

- For sanitation, measures should be aimed at reducing the exposure of residents and visitors to disease risks such as COVID-19 and improving sanitation standards to enhance the visitor experience. Actions should include (i) the adoption of minimum design standards for toilets and solid waste collection facilities at all tour camps in KLNP; (ii) the implementation of procedures for the safe collection, transport, treatment, and disposal of sewage and solid waste; (iii) the establishment of disease risk management procedures by *soum* governments, the KLNP Administration, and tour operators, including routine screening (e.g., temperature checks) of visitors at park entry points, and training to ensure quick and effective responses to contain and treat infected persons in the event that COVID-19 is detected in the future; (iv) the conduct of programs to raise awareness among local communities and visitors regarding sanitation procedures (e.g., handwashing); and (v) the inclusion of these new procedures in the KLNP management plan and *soum* development plans.

- Establish guidelines and performance standards to regulate the issuance, monitoring, and renewal of tour camp licenses. Recommendations for Mongolian protected areas (Spoelder and Batjargal 2013) provide a comprehensive basis for this. Performance standards should include criteria for waste control and local employment.

- Immediately control effluent from tour camps closest to Khuvsgul Lake. Pollution from these camps presents a key threat to Khuvsgul Lake. Addressing this issue is a high priority to build climate resilience. Measures should include (i) the immediate replacement of inadequate toilet and drainage systems to halt the leakage or overflow of sewage or water from washing into the soil or lake; (ii) the prohibition of phosphate-based detergents, soaps, and other cleaning agents; and (iii) strict compliance monitoring of these measures by the KLNP Administration.

- Implement a program for the phased relocation of tour camps that do not comply with minimum legal distances to Khuvsgul Lake. This measure will reduce pollution risks and remove a precedent for future development in similar locations.

- Design climate-resilient infrastructure. The design, and operation and maintenance, of infrastructure (e.g., buildings, roads, sanitation systems) should avoid impact on permafrost and take into account the increased risk of extreme weather (e.g., storms or fire). Measures should include the selective siting of infrastructure and the use of climate-resilient materials for construction.

- Strengthen community-based tourism. This will increase community resilience to climate change through alternative livelihoods and support the improved management of natural resources through participatory planning with the KLNP Administration.

- Improve visitor safety in the face of climate hazards. Medical facilities at KLNP comprise two small, poorly equipped clinics (in Khatgal and Khankh towns). As visitor numbers increase, medical facilities and emergency services should be expanded.

These measures will help strengthen resilience to climate change by reducing the existing impacts from tourism and enhancing support for park management.

Improving Habitat Connectivity

Climate change is likely to cause severe ecological impacts in KLNP through the projected changes and/or the decline of ecosystems (Chapter 5). Plant and animal populations will be forced to adapt to new climate conditions or—if there is sufficient habitat and no barriers to movement—shift northward or to higher elevations to remain within cooler and wetter conditions. At the ecosystem level, a key approach to building climate resilience for biodiversity is to improve habitat connectivity by (i) extending the boundary of a protected area to adjacent lands, if this is feasible and if such lands support similar ecosystems; and (ii) exercising coordinated management over a network of protected areas to provide large corridors of land for ecosystems to shift to, within a range of elevational and latitudinal gradients.

Four protected areas occur within 100 km of KLNP (Figure 1): the Khoridol Saridag Strictly Protected Area (KSSPA; 2,274 km²), the Ulaan Taiga Strictly Protected Area (UTSPA; 4,349 km²), and the Tengis–Shishged National Park (TSNP; 8,690 km²) in Khuvsgul *aimag*; and the Tunkinsky National Park (TNP; 11,837 km²) in the Russian Federation. The KLNP shares boundaries with the KSSPA and the TNP; together, these three protected areas encompass an almost uninterrupted corridor of habitats extending over 260 km from the southwest to the northeast. The north–south orientation of this network is significant, as it increases the opportunity for species to shift northward to cooler and wetter conditions as the climate changes. The potential for transboundary conservation for KLNP is recognized (MacKinnon et al. 2005), and planning is underway to establish a transboundary protected area with the TNP (Whaller 2015).

The proximity of nearby protected areas and the assessment of ecosystems that occur within 50 km of KLNP (Chapter 4) indicate two key features relevant to planning for habitat connectivity: (i) most terrestrial ecosystems in KLNP are also present in nearby lands, including protected areas, providing an opportunity to build climate resilience through coordinated land use management over a large area (northern Khuvsgul *aimag* and adjoining areas of the Russian Federation); and (ii) conversely, this approach will not benefit biodiversity in Khuvsgul Lake, as there are no adjoining lakes further north that fish or other organisms could shift to. For Khuvsgul Lake, this situation emphasizes the importance of addressing the existing threats to the lake in order to build resilience to climate change.

Measures to improve habitat connectivity should include the following:

- Extend the western boundary of KLNP in Renchinlkhumbe *soum*. Most of the KLNP ecosystems are well represented in this *soum* (Terrestrial Ecosystems, Chapter 5). Among the *soums* surrounding KLNP, this *soum* has the highest priority for building ecosystem resilience to climate change through habitat connectivity.
- Assess the potential to establish a transboundary border between KLNP and the TSNP.
- Extend the eastern boundary of KLNP in Tsagaan-Üür *soum*. This will increase protection for three KLNP ecosystems (open forest, forest steppe, and riverine shrub).
- Improve conservation planning in northern Khuvsgul *aimag*—for KLNP, KSSPA, UTSPA, and TSNP, and the lands between them—and with the TNP in the Russian Federation. This region remains relatively undisturbed. The presence of an existing network of protected areas and the threat of climate change emphasize the need for a regional approach to land planning that prioritizes conservation and sustainable development. The KSSPA and the TSNP also support large areas of alpine steppe and alpine tundra—two ecosystems projected to decline in KLNP (Chapters 4 and 5). Under climate change, the regional importance of both sites for the long-term protection of these ecosystems will increase.

Strengthening Park Management

Building resilience to climate change will require institutional reform and expanded technical and financial resources to support the management of KLNP.

At the national level, there is limited policy guidance to manage livelihoods, tourism, waste, or climate change in protected areas. The duration of protected area management plans (generally 4–5 years) is also insufficient for long-term planning to build resilience to climate change. At the site level, the management of KLNP is subject to planning by different provincial and national agencies, with differing jurisdictions and priorities. The KLNP management plan (MEGDT 2014b) is, in principle, the lead document for the park; but other plans—among them, provincial and *soum* development plans and the national Road Development Program, 2017–2021 (Government of Mongolia 2017), which includes road planning for KLNP (Ecological Values, this chapter)—affect the KLNP management and conservation. There is limited integration between these plans. Personnel of the KLNP and Khuvsgul Lake–Eg River Basin administrations, and the *soum* agencies, have limited training and resources to plan and manage conservation, tourism, and waste in the park. In 2019, the KLNP Administration had an annual budget equivalent to only $132,000 and an average of one ranger per 50,000 ha of the park (KLNP Administration, unpublished data). Poor transparency in the issuance of tour camp licenses has resulted in land conflicts and complaints from tourism operators and herders. One *soum*, Khankh, has no development plan, yet is located entirely within KLNP.

The current land use zoning of KLNP also hinders planning for sustainable land management and conservation. Although most of KLNP is designated as a special zone (the highest level of administrative protection), the entire shore of Khuvsgul Lake is zoned for tourism or limited use (Figure 1; Chapter 2). These designations enable tourism and livestock grazing, but have not been based on strategic planning for future tourism development, patterns of herder use for livestock grazing, overgrazing, and hazards from flooding.

These issues are complex and present major challenges for the management of KLNP. Previous reviews have highlighted the urgent need to improve park management and address threats to KLNP arising from unmanaged tourism, livestock overgrazing, and the risks of unsustainable future development (IUCN 2008; MEC 2013; Triepke et al. 2013; Goulden and McIntosh 2018). The government has initiated efforts to address these various issues. These efforts include the revision of Mongolia's Law on Special Protected Areas and, for KLNP, the issuance of a Presidential decree to improve park management and protection, the installation of an entrance gate and fee collection station to regulate public access (in 2017), and the suspension (in 2018) of new tourism licenses pending further review.

Further measures to strengthen the management of KLNP should include the following:

- Establish a national planning mechanism (e.g., a committee) for the northern region of Khuvsgul *aimag*, covering at least KLNP, KSSPA, UTSPA, and TSNP, and adjacent lands. The committee should include all national and local agencies involved in land use planning for the region. Tasks for such a committee should include (i) a strategic environmental impact assessment of planned development; and (ii) a regional plan for conservation and development, with land use planning in the context of climate change.
- Establish a 30-year master plan for KLNP, for the period 2020–2050. To support the vision for KLNP (Chapter 2), the plan should establish long-term objectives, with measurable, time-based targets for conservation, livelihoods, tourism, waste control, and infrastructure development. Assuming that the KLNP management plan will be updated every 4–5 years, the master plan should allocate actions over the next six to eight management plans up to 2050 to achieve these objectives. This long-term approach will align with the national development framework, which covers the same period (Government of Mongolia 2020).
- Establish the KLNP master plan and the KLNP management plan as the lead documents for all land use planning for the park. This should be achieved through policy or legal reform requiring national and local agencies to ensure that other national and provincial documents are consistent with, and incorporated

in, these plans. This measure will support multisector planning and consistency across ministries and departments.

- Strengthen provincial- and *soum*-level land use planning for KLNP by (i) preparing a *soum* development plan for Khankh *soum* (this *soum* encompasses the largest area of KLNP, yet has no development plan; and planning is urgently required to support conservation and livelihoods); and (ii) revising the development plans for Alag–Erdene, Chandmani–Undur, Renchinlkhumbe, and Tsagaan–Üür *soums*, and Khatgal town to improve alignment with the KLNP management plan.

- Review the effectiveness of the current management of KLNP, using a standardized methodology (e.g., the management effectiveness tracking tool; Stolton et al. 2019). This will help guide the updating of the KLNP management plan.

- Update the KLNP management plan, for the period 2020–2024. The updated plan should include (i) subplans for biodiversity conservation, livelihoods, tourism, sanitation and waste management, and road and traffic development; (ii) maps of key features for management, e.g., the locations of tour camps, delineated grazing lands, ranger patrol routes, and park zoning; (iii) roles of all relevant agencies (national and local) in park management; (iv) cross-referencing to all other relevant planning documents; and (v) financial and technical resources required to implement the plan. These revisions will provide, for the first time, an integrated and multisector planning document to support the KLNP Administration and other agencies.

- Revise the park zoning to strengthen land planning for conservation, livelihoods, and tourism under climate change, particularly (i) to expand the special zone to increase protection for six ecosystems that are already restricted in extent and/or under threat, and will be further impacted by climate change: alpine steppe, alpine tundra, meadow, riverine meadow, riverine shrub, and Khuvsgul Lake (Chapter 5); and (ii) for Khuvsgul Lake, which is entirely zoned for tourism (Park Management, Chapter 5), to rezone most of the lake as a special zone to help reduce the risk of overdevelopment, preserve the lake's scenic values, and protect fish spawning sites (Goulden and McIntosh 2018). Three areas of the lake already developed for tourism (the northern, southwestern, and southeastern shores; Figure 1) should be maintained as a tourism zone.

- Establish a long-term monitoring program, up to at least 2050, to measure progress toward management targets under the proposed new KLNP master plan and updated management plan. The program should include indicators for biodiversity, livelihoods, tourism, lake pollution, and management effectiveness. The methodology should be replicable, quantifiable, and as simple and cost-effective as possible. Measurement of indicators should include science-led and participatory approaches; the latter will be especially important for stakeholder ownership of monitoring results and corrective actions to address identified issues. The current study contributes toward such a program, as it provides a publicly available baseline of bioclimatic zones and ecosystems in KLNP under baseline and future climate scenarios.

- Support continued scientific research in KLNP. Research areas to be prioritized should include the distribution and mapping of permafrost, vegetation communities, and threatened species of plants and animals. This information will help conservation and land use planning, e.g., to avoid infrastructure damage to permafrost.

- Strengthen technical and financial resources for management. Measures should build on previous efforts by the government, civil society organizations (CSOs), and development agencies (Chapter 2) and include (i) technical training in conservation, tourism, and sanitation and waste management; and (ii) sustainable financing mechanisms.

- Support management planning and implementation that are participatory and inclusive. In particular, plans for conservation, tourism, and waste management should be developed and implemented with local communities through gender-inclusive approaches that also ensure the inclusion of the poor and vulnerable households.

These various measures will contribute to the long-term management of KLNP that is holistic in scope, adaptive, includes climate change within planning, and employs science-led and participatory approaches. The range of interdisciplinary measures needed illustrates the complexity of protected area management and the added challenges due to climate change.

7 Role of the Asian Development Bank

Key Messages

- ADB's support to improve livelihoods and develop tourism sustainably at Khuvsgul Lake National Park (KLNP) is being provided under two projects, covering the period 2016–2024. These are the first projects to adopt a multisector planning approach at KLNP.
- The ADB projects build on a large platform of previous support for scientific research, livelihoods, and park management at KLNP provided by national and international agencies. Collectively, these efforts, especially those addressing livestock and pasture management, pollution, and park management, are helping to build resilience to climate change at KLNP.
- Achieving resilience to climate change for a large protected area with multiple-use objectives and development pressures is beyond the scope of any single agency or project. For KLNP, interagency collaboration and long-term support for biodiversity conservation, livelihoods, tourism, and park management, designed to address site-specific climate impacts, are required to build resilience to climate change.

Country Partnership Strategy

ADB's Mongolia country program (initiated in 1991)[1] includes the provision of technical and financial support for Khuvsgul Lake National Park (KLNP) under two projects implemented between 2016 and 2024: (i) the Integrated Livelihoods Improvement and Sustainable Tourism in Khuvsgul Lake National Park Project, and (ii) the Sustainable Tourism Development Project. The country program is guided by a country partnership strategy, which identifies programmatic goals and approaches and is periodically updated. For the period 2017–2020, the strategy is focused on three priorities: promoting economic diversification and job creation, increasing the inclusiveness of social service delivery, and strengthening environmental sustainability (ADB 2017). A new strategy, for the period 2021–2024, is under preparation. Both strategies emphasize the importance of support for natural resources management, climate change adaptation and mitigation, and the development of sustainable tourism. Preparation for ADB project assistance for KLNP began in 2014, at the request of the Ministry of Environment and Tourism to help develop sustainable tourism, livelihoods, and pollution control. The two projects for KLNP are ADB's first projects in Mongolia to focus on tourism and protected area management.

Project Support, 2016–2024

The Integrated Livelihoods Improvement and Sustainable Tourism in Khuvsgul Lake National Park Project was implemented between January 2016 and June 2020 and was funded by a grant of $3.0 million (ADB 2015). This is being followed by the Sustainable Tourism Development Project, which began in August 2019 and is scheduled for completion by December 2024, and which is funded by loans totaling $38.0 million (ADB 2019). The first project focused on KLNP; the second is being implemented at KLNP as well as at another national park in a different *aimag* (ADB 2019).

The scope and design of each project were based on assessment of the threats and management issues facing KLNP; consultations with stakeholders (government agencies, communities, tour operators, CSOs, development agencies); the importance of aligning with national regulations for protected areas and the KLNP management

---

[1] Information about the ADB Mongolia country program is available at https://www.adb.org/countries/mongolia/overview.

plan; and ADB requirements for designs to be technically and financially viable, inclusive, gender-equitable, climate resilient, and safeguard-compliant. Support provided to KLNP by other development agencies as well as CSOs was reviewed to assess the potential for ADB to add value. The grant-funded project was supported by a donor (see Acknowledgments), and designs were also required to meet funding guidelines (i.e., a focus on livelihoods and nonstructural measures). The loan-funded project includes structural components and, for these, designs were guided by projections of tourism and population growth, durability, the potential impacts of climate change on the planned infrastructure, costs of construction and operation and maintenance, and potential project impacts on environmental and social values.

These assessments confirmed the potential for ADB support to add value, through a holistic, multisector approach to livelihoods, tourism, and conservation in KLNP. The following areas were identified for ADB support: institutional strengthening for conservation and tourism, livelihoods, tourism planning and management, waste management, and park management (Bezuijen 2019). Under the Integrated Livelihoods Improvement and Sustainable Tourism in Khuvsgul Lake National Park Project, efforts focused on livelihoods and initial measures for institutional strengthening, waste management, and park management. Under the Sustainable Tourism Development Project, these works are being expanded to encompass all of the identified areas.

The Integrated Livelihoods Improvement and Sustainable Tourism in Khuvsgul Lake National Park Project was designed with the following target outcome: *livelihoods* and *sustainable tourism in the five soums of KLNP improved and integrated*. Three outputs were designed to achieve this outcome (ADB 2015): community-based tourism in Khatgal and Khankh settlements promoted (output 1), capacity for sustainable livestock and pasture management in KLNP and buffer zone improved (output 2), and waste management around Khuvsgul Lake strengthened (output 3). Under output 1, a park management council (KLNP's first mechanism for participatory dialogue with communities and tour operators), codes of conduct for tour operators and visitors (to raise awareness of park regulations), and a microfinancing scheme (to provide low-interest household loans for enterprises compatible with the KLNP regulations) (Nergui 2020) were established. Under output 2, herder groups, pasture management plans, and community pasture monitoring were established for grazing lands; and herders were trained in pasture management. Under output 3, public toilets and litter bins were installed at tourism locations, and community teams were mobilized to manage these (WaSH Action of Mongolia 2020); a water quality monitoring program for Khuvsgul Lake was established (Natural Sustainable 2019); and a review of park zoning was conducted (Enkhtaivan and Munguntulga 2019).

Project implementation was coordinated by a project implementation unit under the Ministry of Environment and Tourism, which worked closely with project stakeholders and ADB. Operational and financial sustainability for the mechanisms established under the project was an overarching project goal. During the project, the designs and procedures for operation and maintenance of the project mechanisms were piloted, refined, and then transferred (with technical and training support) to the work plans of local agencies. Prior to project completion in 2020, the project mechanisms were being managed independently by these agencies: the KLNP Administration (for the KLNP management council, codes of conduct, pasture management plans, and coordination of herder groups); *soum* government agencies (for toilets, litter bins, and community waste management teams); a CSO (ecoLeap, for waste management); National Agency for Meteorology, Hydrology and Environmental Monitoring (for water quality monitoring program); and the State Bank of Mongolia (for the microfinancing scheme).

Most of these measures are the first inclusive and sustained mechanisms for livelihoods, pasture management, pollution control, water quality monitoring, and park management for KLNP. About 3,300 residents (54% of the KLNP population and 16% of the total population of five *soums* in 2018; Chapter 2) benefited from the project through the issuance of household loans, employment, or training (MET 2020).[2]

---

[2]   In accordance with ADB procedures, a full review of project results will be conducted 1 year after completion (in 2021) and published on the ADB website (www.adb.org). The project progress reports (2016–2020) and other cited project documents are available on the ADB website (see links to the cited project documents in the References). Results of the water quality monitoring program are available at http://water.sain.site/ and http://www.tsag-agaar.gov.mn/.

The Sustainable Tourism Development Project was designed to build on the grant-funded project. The target outcome is *sustainable and inclusive tourism in Khuvsgul Lake National Park and Onon-Balj National Park (in Khentii aimag) developed*. To achieve this outcome, four outputs have been designed (ADB 2019): inclusive planning and capacity for community-based tourism enhanced (output 1), enabling infrastructure for tourism constructed (output 2), waste management improved (output 3), and park management strengthened (output 4). Under output 1, the first development plan for Khankh *soum* will be prepared and the existing plan for Khatgal town revised (to improve planning in the context of a protected area); procedures for the management of tour operations and a tourism certification scheme will be established (to embed social and gender targets and environmental standards in the licensing process); links between household enterprises in KLNP and provincial markets will be strengthened; and the capacity building initiated under the grant-funded project will be extended. Under output 2, the project will upgrade unsealed roads along the heavily visited southwestern shore of Khuvsgul Lake and establish small car parks and barrier gates to regulate public and vehicle access, improve safety, and reduce dust inputs to Khuvsgul Lake and damage to the lake shoreline.

Under output 3, the project will assist tour camps to adopt the toilet designs piloted under the grant-funded project and also install these toilet designs in public campsites and tourist venues, tailored to local conditions. Wastewater treatment plants will be constructed at Khatgal and Khankh towns, and two existing landfills at each town will be upgraded. Under output 4, a new KLNP headquarters and visitor center, two fee collection stations (near Khatgal and Khankh towns), and a road control station (along the southwestern shore of Khuvsgul Lake, to restrict further development along the western shore of the lake) will be constructed. About 15 km of public trails, subject to high seasonal use, will be rehabilitated. The KLNP management plan will be revised to include, for the first time, measures for livelihoods, tourism, and waste control; an updated operational budget; and amendments to the park zoning. Training will be conducted for park staff and *soum* government agencies to implement the revised KLNP management plan, particularly for tourism planning and waste management.

These designs are intended to strengthen conservation and development planning for KLNP. Quantitative targets include improved land use planning for the two largest settlements in KLNP (Khatgal and Khankh towns; over 6,000 residents), improved management of tour camps, greater protection for over 40 km of shoreline of Khuvsgul Lake in the most intensively visited region of KLNP, and design capacity for waste management to meet projected demand for at least the first 15 years of operation (ADB 2019). The designs support the objectives of the KLNP management plan and contribute to the goals of the National Biodiversity Program, 2015–2025 (for improved protected area infrastructure, management capacity, and protection of threatened species) (Government of Mongolia 2015). A key outcome-level project target is that, by 2025, KLNP is nominated for inclusion in the Green List of Protected and Conserved Areas (IUCN 2020), a global standard for protected area management. This is an ambitious target—no protected areas in Mongolia have yet been listed, and relatively few globally (IUCN 2020). If achieved, this would indicate that KLNP has met best-practice criteria for conservation, social benefits, sustainability, and governance.

Contributing to Climate Change Resilience

In the context of climate change, the ADB projects contribute to many of the actions needed to build climate resilience in KLNP. Of the three approaches to building resilience to climate change in protected areas—i.e., address existing threats, improve habitat connectivity, and strengthen park management (Chapter 6)—the ADB projects contribute to two of these: addressing existing threats and strengthening park management.

Addressing existing threats. Of the seven key existing threats to ecological values in KLNP (Ecological Values, Chapter 6), the ADB projects are helping to address the two most severe—livestock overgrazing and unmanaged tourism. For livestock overgrazing, the formation of the KLNP management council, herder groups, pasture management plans, and community-based pasture monitoring (under the Integrated Livelihoods Improvement and Sustainable Tourism in Khuvsgul Lake National Park Project) has contributed to improved management of the three most intensively grazed ecosystems in the park (open forest, forest steppe, and lower mountain steppe; Livelihoods, Chapter 5). Results relevant to building resilience to climate change include the mapping of pasture condition for over 220,000 ha of grazed lands, and the reestablishment of seasonal rotation for some grazing lands around Khuvsgul Lake to reduce grazing pressure (MET 2020). These measures included community-based pasture mapping led by technical specialists using a replicable methodology (quadrat-based sampling of vegetation cover, quality, and species richness) and consultations between herders and the KLNP Administration.

For Khuvsgul Lake, the waste management components of the ADB project are helping to address water pollution, the greatest threat to the lake's ecology under climate change (Khuvsgul Lake, Chapter 5). The various project approaches to improving toilet design standards and procedures and facilities for waste collection, treatment, and disposal, combined with improved local management capacity, are intended to reduce waste inputs to Khuvsgul Lake. The planned road upgrades will reduce the input of dust to Khuvsgul Lake, a localized but important source of lake pollution (Tourism, Chapter 6). The water quality monitoring program developed under the grant-funded project will help measure the lake's "health" over time; this will be especially important under climate change, for timely detection of increased nutrient levels and/or algal blooms as water temperatures rise. This also complements regional efforts to develop a transboundary Baikal Basin Water Quality Monitoring Program (Whaller 2015).

The livelihood improvement components of the ADB project will strengthen the resilience of local communities to climate change through income diversification, improved food security, and improved land management. These include project measures for herding livelihoods (improved sustainability of grazing; pasture health; preparation of winter fodder; and improved quality of wool, meat, and dairy products) and tourism (improved skills, and provision of local goods and services). The improved toilet designs and facilities for treatment and disposal of sewage and solid waste will also strengthen the safety of residents and visitors under climate change, by reducing the risk of disease spread as conditions become warmer. The measures to improve road and traffic management will help address the risk of increased hazards to residents and visitors and infrastructure resilience to increased storm intensities.

Strengthening park management. At the institutional level, the measures to be conducted under the Sustainable Tourism Development Project of updating the KLNP management plan, park zoning, and *soum* development plans, and building governance capacity will strengthen park management under climate change. The present study and the review of park zoning (Enkhtaivan and Munguntulga 2019), under the grant-funded project, provide new recommendations for the KLNP zoning based on science and participatory approaches. Land use planning will be aligned between the KLNP management plan and the development plans for Khatgal town and Khankh *soum*, and the herder management plans will be integrated into the KLNP management plan. The improved cooperation between the KLNP Administration, communities, and tourism operators will also increase the ability of the government to monitor and evaluate the results of management decisions.

Climate-resilient infrastructure designs. Both ADB projects were designed to minimize impacts on the climate, as well as adverse impacts from climate change that might affect the project. For the Integrated Livelihoods Improvement and Sustainable Tourism in Khuvsgul Lake National Park Project, toilet systems were designed to be non-flushing (to save water), sited away from Khuvsgul Lake, and installed on elevated concrete bases to avoid soil excavation and damage to permafrost. For the microfinancing scheme, loan approvals are linked with environmental and social criteria to ensure compliance with park regulations—e.g., the exclusion of financing support for activities involving goat husbandry because of the high grazing impacts of goats (Livelihoods, Chapter 6). For the Sustainable Tourism Development Project, the locations for the planned KLNP headquarters and visitor center, wastewater treatment plants, and fee collection and road control stations were selected to avoid sites subject to subsidence from melting permafrost and to minimize damage to permafrost and vegetation. The road designs include insulated road bases to strengthen cold tolerance and protect permafrost from vehicle damage, the wastewater treatment plants apply designs proven to operate in cold climates, and the landfill designs include clay lining to avoid soil and water pollution.

Applying adaptive management to project design. The Sustainable Tourism Development Project is in the early stages of implementation, and the findings of the present study provide new opportunities to help strengthen the project design. The findings also help highlight the importance of an adaptive approach to project management by responding to new information and changing conditions. The ADB projects were not designed to be exhaustive in scope: they focus principally on measures to improve livelihoods and achieve sustainability in tourism. They include most, but not all, of the recommended measures to build resilience to climate change (Chapter 6). The Integrated Livelihoods Improvement and Sustainable Tourism in Khuvsgul Lake National Park Project included measures to help address livestock and pasture management, but there is less focus on these issues under the Sustainable Tourism Development Project. Both projects were also designed prior to the outbreak of COVID-19, and the projections of tourism growth applied to the project designs were based on pre-COVID-19 growth scenarios.

The study findings indicate the need for a greater project focus on (i) measures to reduce pollution of Khuvsgul Lake from livestock, given the sensitivity of the lake to water pollution under climate change; and (ii) the inclusion of measures for transboundary management planning in the KLNP management plan to better protect ecosystems as their distributions shift under climate change. The outbreak of COVID-19 necessitates a greater focus on sanitation to help manage and reduce the risks of future disease outbreaks, and new planning to support local agencies and communities with post-COVID-19 tourism recovery in KLNP.

It will also be essential to maintain and scale up collaboration with CSOs and other agencies established under the grant-funded project. ADB's project support complements, and cannot replace, the invaluable role of CSOs and other agencies that provide support for livelihoods and conservation in KLNP (Chapter 2). To maximize the benefits of ADB support, collaboration and coordination with agencies focused on parallel initiatives for biodiversity conservation (e.g., research, surveys, and actions to address hunting and overfishing) and livestock and pasture management will be especially important to help achieve climate-resilient conservation and development at KLNP.

Appendix: Methodology

1. **Compilation of weather data for Khuvsgul Lake National Park.** Data for three parameters (air temperature, water temperature of Khuvsgul Lake, and precipitation) were provided by the National Agency for Meteorology, Hydrology and Environmental Monitoring. These data came from two meteorological stations, in Khatgal and Khankh towns, for the periods 1963–2016 (Khatgal town) and 1985–2016 (Khankh town). Data were analyzed for annual and seasonal trends. Statistical significance was measured using the Mann–Kendall test (significance level $p < 0.05$), and the slope of the linear trend was estimated with the nonparametric Sen's method (Gilbert 1987).

2. **Spatial data.** Modeling was conducted using the geographic information systems, ArcGIS 10.5 and QGIS. Spatial files of the Khuvsgul Lake National Park (KLNP) boundary, Khuvsgul Lake, roads, management zones, and locations of tour camps and herder camps (created in 2011 by D. Enkhtaivan from the digitizing of 1:100,000 topographic maps; accuracy ±50 meters) were provided by the KLNP Administration. Coordinates of the KLNP boundary and management zones are given in the KLNP management plan (MEGDT 2014b) in degrees, minutes, and seconds (to 3 decimal points and an accuracy of ±30 centimeters for latitude and ±70 centimeters for longitude). The spatial accuracy of camp locations, including the locations of new camps, was confirmed by the KLNP staff using a handheld geographic positioning system.

Spatial files for roads, rivers, *soum* (district) administrative boundaries, and the international border were downloaded from the open-source website of the Administration for Land Affairs, Geodesy and Cartography (2015). No data on spatial accuracy is provided with the files; it is assumed they were digitized from 1:100,000 maps (accuracy ±50 meters). For climate data and elevations, GeoTIFF spatial files (30-meter resolution) were downloaded, respectively, from the open-source websites of WorldClim and Advanced Land Observing Satellite–2 websites (JAXA/EORC 2020) and used in compliance with the terms of public use. Due to the large pixel sizes of some data, study findings for spatial areas are given in square kilometers (km^2). The use of some map layers with unknown accuracy, combined with rounding to the nearest km^2, may explain any discrepancies between the areas presented in this study and other studies. Such discrepancies are likely to be small.

3. **Development of a model of climate change for the Khuvsgul Lake National Park.** To develop a model of the current climate of KLNP, temperature and precipitation data were downloaded from an open-source model, WorldClim v.1.4 (Hijmans 2015; Hijmans et al. 2005). The WorldClim model divides the world's surface into a grid of cells, each 1 km^2 in size, of mean monthly climate variables for the period 1960–1990. It is based on a large number of global records of precipitation and temperature. Sources for the data used in the study include weather stations in northern Mongolia and nearby areas of the Russian Federation. Three other climate parameters were derived from the WorldClim data: potential evapotranspiration (PET), aridity-wetness index (AWI), and growing degree days (GDD) (see Glossary). The five climate parameters were used to establish the climate conditions in KLNP for the period 1960–1990. This represented the "baseline climate" for the study. The methodology is in Zomer et al. (2008) and Zomer et al. (2014).

The WorldClim website also provides projections of climate change from global climate models (GCMs) obtained from the Coupled Model Intercomparison Project–Phase 5 (CIMP5; Meehl and Bony 2011) that have been downscaled and bias-corrected to a spatial resolution of 30 seconds of a degree of latitude and longitude (about 900 meters at the equator). These projections comprise four scenarios of changes in the levels of greenhouse gas concentrations, termed representative concentration pathways (RCPs), for two time periods: 2050 (average for 2041–2060) and 2070 (average for 2061–2080). An analysis of the GCMs was conducted to identify the models best demonstrated to predict the baseline climate in the study area. Seventeen downscaled models were compared to identify the "risk space" projected by the set of available GCMs for the study area (see figure, p. 52).

Selection of the GCMs for the study was based on (i) proximity to the study area (GCMs parameterized, calibrated, or evaluated using meteorological data of the People's Republic of China and the Russian Federation); and (ii) GCMs best demonstrated to accurately predict two global climatic processes, which are the main drivers of Mongolia's climate (especially rainfall)—the Asian monsoon (Davi et al. 2010) and El Niño events (which appear to influence the Asian monsoon based on changes in sea surface temperature) (Fan et al. 2017).

Analysis of GCM performance, as per best practices of the Intergovernmental Panel on Climate Change (IPCC) for multimodel ensembles (Knutti et al. 2010), indicates that using the mean value for the various available climatic variables (e.g., mean monthly precipitation or mean monthly maximum temperature) of an ensemble of the top performing models can improve historical accuracy and decrease variability associated with individual model performance. Agreement among models was used as a measure of variability, with agreement among environmental strata classes (based on a majority classification, per each grid cell) used as a measure of uncertainty (see point 5, p. 54). The figure below highlights the variability among the models, with most clustering around the average, and how two outlier models (GF85 and MI85) project larger changes under the RCP8.5 scenario (outer edges of the polygons).

Based on this analysis, data from 10 GCMs available from the WorldClim website were selected (see table, p. 53), downscaled, and bias-corrected. Three GCMs found to perform well in Mongolia (the MPI models from the Max Planck Institute for Meteorology, and EC_EARTH model) were not available from the WorldClim website. Instead, three other well-performing GCMs were included: two regional models, BCC-CSM1-1 (BC) from the People's Republic of China and INMCM4 (IN) from the Russian Federation; and one model from the Norwegian Climate Centre, NorESM1-ME (NO). In addition, a National Oceanic and Atmospheric Administration (NOAA) model was replaced with the model GISS-E2-R (GS) from the National Aeronautics and Space Administration (NASA) Goddard Institute for Space Studies. Two models, MIROC5 (MI) and GFDL-CM3 (GF) consistently projected

Comparison of 17 Downscaled CIMP5 Earth System Models (Meehl and Bony 2011) and WorldClim (Hijmans 2015; Hijmans et al. 2005) for Khuvsgul Lake National Park (RCP8.5)

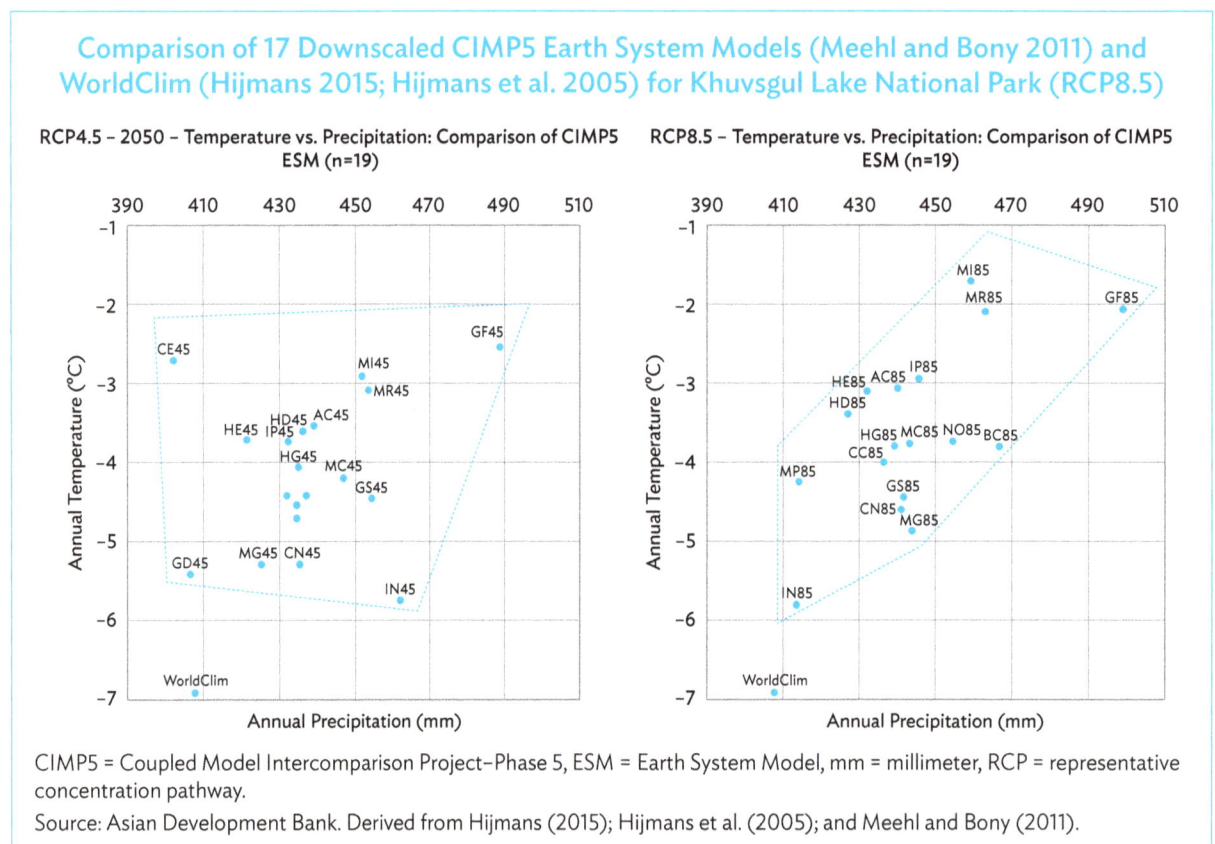

CIMP5 = Coupled Model Intercomparison Project–Phase 5, ESM = Earth System Model, mm = millimeter, RCP = representative concentration pathway.

Source: Asian Development Bank. Derived from Hijmans (2015); Hijmans et al. (2005); and Meehl and Bony (2011).

Models Selected for the Climate Modeling Undertaken for This Study

Model	Code	Agency
BCC-CSM1-1	BC	Beijing Climate Center, People's Republic of China
CCSM4	CC	National Center for Atmospheric Research, United States
CNRM-CM5	CN	Centro Euro-Mediterraneo per I Cambiamenti Climatici, Italy
GFDL-CM3	GF	NOAA Geophysical Fluid Dynamics Laboratory, United States
GISS-E2-R	GS	NASA Goddard Institute for Space Studies, United States
HadGEM2-CC	HG	Met Office Hadley Centre, United Kingdom
INMCM4	IN	Institute for Numerical Mathematics, Russian Federation
MIROC5	MI	MIROC, Japan
MRI-CGCM3	MG	Meteorological Research Institute, Japan
NorESM1-ME	NO	Norwegian Climate Centre, Norway

MIROC = Japan Agency for Marine-Earth Science and Technology, NASA = National Aeronautics and Space Administration, NOAA = National Oceanic and Atmospheric Administration.

Source: Government of Switzerland, Federal Roads Office.

higher temperature and precipitation changes under both RCP scenarios (see figure, p.52); both were also included in the study analysis to illustrate the potential extent of climate change in KLNP under high-impact scenarios.

4.   **Identification of bioclimatic zones and environmental strata in the Khuvsgul Lake National Park.** A global environmental stratification (GEnS) (Metzger et al. 2013) was used as basis for the study. This model is a statistical stratification of the world's land surface into 125 environmental strata, aggregated in 18 environmental zones, ranging from "A" (arctic) to "R" (extremely hot and moist), based on high-resolution global climate data sets (Metzger et al. 2013). The GEnS (i) quantitatively relates the spatial distribution of ecosystems to an identified set of bioclimatic parameters; (ii) provides a consistent methodology across landscapes and countries that have so far mostly been studied using different protocols, approaches, and taxonomies; and (iii) allows for a statistical modeling of bioclimatic zonal shifts that can be used to estimate the direction and magnitude of impacts on ecosystems due to climatic changes. Metzger et al. (2013) identified that 99.9% of variation between the zones and strata is determined by four variables: GDD, AWI, and PET (estimated from the WorldClim data); and monthly mean temperature seasonality (available from the WorldClim website). For the study, these four variables were used as inputs to the ISODATA clustering routine in ArcGIS to classify the GEnS classes (Metzger et al. 2013).

The study identified three zones and nine strata in KLNP for 1960–1990 (Chapter 4). The strata are represented by numbers, with advancing numbers representing decreasing elevation, decreasing wetness, and increasing temperature. The zones are summarized as follows:

- **Zone D (extremely cold and wet).** This zone has the lowest minimum temperature and (together with Zone F) the least GDD per year (<1,000) of the three zones. Annual precipitation is the highest in KLNP. This, combined with low PET, produces the highest AWI (>1.5). Low temperatures and GDD indicate that biological processes are limited by temperature (except in peak summer). High precipitation and AWI indicate that moisture is abundant and biological processes are not limited by dryness.
- **Zone G (cold and mesic).** This zone has the most GDD per year (1,000–2,500) and highest temperatures, i.e., biological processes operate more effectively than in Zone D and over a longer season. It has the lowest mean annual precipitation and highest PET, producing the lowest AWI (0.6–1.0) and driest conditions of the three zones. Most of the year, moisture is adequate for growth, but is limited in summer.
- **Zone F (extremely cold).** This zone has <1,000 GDD per year and an AWI between 0.6 and 1.0. There are seven strata in this zone in KLNP (Chapter 4). These show characteristics between zones D and G. Higher code numbers for strata generally indicate a trend in each variable, e.g., temperature, which increases from

stratum F_03 (lowest) to F_13 (highest). Four strata (F_03, F_05, F_06, and F_07) are colder than the other strata, have fewer GDD, lower PET, and variable rainfall. This leads to a range of AWI values and limits biological processes, except in summer. Three strata (F_08, F_10, and F_13) are warmer and have more GDD, so biological productivity is higher. Stratum F_08 has high precipitation, low moisture stress, and high GDD—i.e., biologically active. Stratum F_10 (the most widespread) and F_13 have low AWI—i.e., experience periods of moisture stress.

5.   **Modeling of future bioclimatic conditions.** The identified set of statistically significant climate parameters and the statistical profiles of the bioclimatic zones and strata of the KLNP environmental stratification of current climatic conditions (averaged 1960–1990) were used to reconstruct the bioclimatic stratification based on projected future conditions from the 10 selected GCMs. The statistical signature profiles of the environmental strata of KLNP identified for the baseline period were reconstructed based upon a multivariate analysis (maximum likelihood classification) of the five bioclimatic variables using the maximum likelihood classification algorithm in ArcGIS 10.5. The profiles were then used to project the future spatial distribution of the GEnS classes based upon the modeled future climate conditions in 2050. All models in each RCP were combined into a majority ensemble result, using the class with the majority of occurrence within any particular grid cell as the class for that location. The strata continue to represent bioclimatic conditions similar to the original strata (i.e., recent climatic conditions) but may shift in areal extent or location. The change in distribution of the strata was analyzed and used as a surrogate measure to describe the potential projected macro-level impacts of climate change on terrestrial ecosystems (Metzger et al. 2013; Zomer et al. 2008; Zomer et al. 2014).

Uncertainty analysis. Within the environmental strata analysis, the results for each of the 10 models were evaluated for each grid cell, and the class (environmental strata) with the majority of occurrence was used as the class for that location. The rate of occurrence of other classes was used as a measure of the uncertainty among models. Within KLNP, there was generally a high degree of agreement among the models used in the 10-model ensemble analysis, sufficient to say that the results represent a "best guess" consensus among the models. The percentage agreement is 72% for RCP4.5 and 71% for RCP8.5.

Assumptions, sources of uncertainty, and other limitations. Significant limitations are associated with the models, downscaling and interpolation of weather data and GCM results, and assumptions in the models developed for this study (Zomer et al. 2014). Although the analysis was conducted at a resolution of 1 km^2, it cannot be expected that any particular grid cell will be correctly classified and accurately predicted when compared with ground-based data. However, in terms of mapping the spatial distribution of a fairly homogenous set of similar bioclimatic conditions, and then projecting these strata into the future by mapping the spatial redistribution of these sets of unique homogenous conditions under future projected climate conditions, these results present an accounting of the current state of the projections for changes in climatic variables for KLNP by 2050. These results allow for an integrative and holistic interpretation of climatic change to be visualized and interpreted quantitatively—i.e., they provide a basis for understanding and assessing trends and directions of change. Other sources of uncertainty include the difficulties associated with model predictions in heterogeneous terrain, such as the mountainous areas of KLNP. Modeling of climate impacts linked with permafrost, a key parameter that influences local climate in Mongolia, was not possible because of a lack of detailed information on permafrost zones in KLNP and inability to model the coupling of soil and air temperatures. Similarly, the bias corrections employed in deriving the WorldClim datasets are based on elevation and are unable to incorporate the formation of temperature inversions that are known to occur in the mountain valleys of KLNP.

6.   **Identification of ecosystems in Khuvsgul Lake National Park.** A terrestrial ecosystem classification developed for northern Mongolia (Heiner et al. 2017) was applied to identify the ecosystems of KLNP. The classification in Heiner et al. (2017) is based on regionally available data sets for elevation, surface hydrology, topography, vegetation, and other variables derived from satellite imagery and a wide range of studies. The classification comprises 99 categories of ecosystems. These represent the full range of living (vegetation)

and nonliving (physical) environment and ecological processes in northern Mongolia. For this study, the raw spatial data in Heiner et al. (2017) were provided by The Nature Conservancy (as an ESRI raster grid file with 60 meter pixels) and used with permission to derive the ecosystem categories for KLNP. Of the 99 categories in Heiner et al. (2017), a total of 15 categories were identified for KLNP. These were aggregated into seven categories, which could be identified and analyzed with confidence from satellite imagery. Summary statistics on the area and distribution of each category were derived. Modeling was applied to the 15 categories as well as the seven aggregated categories.

Khuvsgul Lake was considered as a single ecosystem for this study. The lake supports a diversity of inshore and deep water aquatic habitats, but most of the lake is dominated by deep water. A finer-level analysis of aquatic ecosystems was beyond the study scope.

The extent to which the ecosystems in KLNP occur in surrounding lands was assessed by measuring the area of each category (i) within 50 kilometers (km) of KLNP (in 10 km increments, extending from 20 km up to 50 km from the park boundaries); and (ii) within the surrounding lands of four *soums* (Chapter 2). This encompassed a total area of about 65,000 km^2 north of latitude 49.8° N and west of longitude 102.7° E.

7.   **Identification of climate change impacts on biodiversity.** The environmental strata under the "current" climate conditions (WorldClim Baseline-1990) were overlaid on the derived ecosystem categories from Heiner et al. (2017). Overlaps between the environmental strata and the ecosystem categories under the future climate conditions were compared against the current conditions. Ecosystems were used as a surrogate indicator for biodiversity values, as changes in ecosystems imply direct changes in the living conditions and physical environment of flora and fauna. As with the assumptions incorporated into the bioclimatic modeling, the development of the ecosystem categories of Heiner et al. (2017) was based on data from a range of sources; therefore, it cannot be expected that any particular grid cell will be correctly classified when compared with ground-based data. There are also some overlaps between the definition of the different ecosystems (particularly within the three alpine classes), between the categories of forest ecosystems, and between the categories of steppe ecosystems. The spatial datasets for the study are stored at the Asian Development Bank Data Library (https://data.adb.org/).

8.   **Stakeholder surveys.** Interviews were held with government personnel, tour operators, residents, and tourists in KLNP to document perceptions about climate change and impacts on livelihoods and tourism. Survey questions were designed by the authors. Interviews were conducted between June and August 2018 during the preparation of the Sustainable Tourism Development Project (ADB 2019). Interviews were conducted by the National University of Mongolia (Population Training and Research Center) (covering 700 households, 700 tourists, and 51 tour camp owners) and the authors (covering 33 local government agencies, civil society organizations, tour camp operators, and residents). Interviews by the National University of Mongolia were conducted according to a structured sampling approach to cover the main settlements and tour camps in KLNP; interviews by the authors were conducted opportunistically in the course of other field work. Questions included (i) whether climate change is an issue for livelihoods and/or tourism and the nature of such impacts (e.g., water shortage, livestock disease, declining natural resources); and (ii) whether climate change presents a future threat to livelihoods, tourism, and/or natural resources. Tour operators were asked to rank how much climate change affected their profit.

9.   **Identification of climate change impacts on livelihoods, tourism, and park management.** In addition to impacts on biodiversity, the study results were used to identify the risks of climate change to livelihoods and tourism in KLNP and the implications for park management. Support for livelihoods and tourism development are key priorities for the management of KLNP, in addition to biodiversity conservation (MEGDT 2014b), and a holistic approach to these different dimensions was adopted for the study. Preliminary measures to help build resilience to climate change were identified for biodiversity conservation, livelihoods, tourism, and park management.

Glossary of Technical Terms

Adaptation	The process of adjustment to actual or expected climate and its effects in order to either lessen or avoid harm or exploit beneficial opportunities (IPCC 2014, 76).
Aridity–wetness index	A measure of the moisture (precipitation) available for plant growth. The higher the value, the more moisture is available. It is based on a ratio of the amount of annual precipitation compared with the potential evapotranspiration.
Bioclimatic stratification	Division of a region into areas (strata) based on climatic conditions that influence plant growth. Strata are derived through a statistical clustering of four parameters: growing degree days, aridity–wetness index, monthly mean temperature, and potential evapotranspiration. Similar strata are grouped into zones.
Bioclimatic stratum	The smallest division of an area that is produced as part of a bioclimatic stratification. An area that experiences similar bioclimatic conditions.
Bioclimatic zone	A grouping of bioclimatic strata that share similar and biologically important climate measures. Derived through the statistical clustering of climatic variables.
Climate change	A change in climate attributed directly or indirectly to human activity that alters the composition of the global atmosphere, and occurring in addition to natural climate variability observed over comparable time periods (United Nations Framework Convention on Climate Change).
Ecosystem	A type of natural habitat that comprises a living component (vegetation, animals) and a nonliving component (physical environment and processes). Ecosystems occur at distinct spatial scales and in patterns driven by underlying physical processes (adapted from Heiner et al. 2017, 22 and 110).
Growing degree days	A measure of the annual seasonal duration available for plant growth, defined as the accumulation of days where mean daily temperature is above 0°C.
Mesic	A climate receiving a moderate amount of precipitation, where moisture supply is balanced between input and evaporative losses.
Permafrost	Ground (soil or rock and included ice or organic material) that remains at or below 0°C for at least 2 consecutive years (IPA 2015).
Potential evapotranspiration	A measure of the total potential amount of transpiration (from plants) and evaporation (from soil and plant canopy) that can be expected under existing or projected temperature and relative humidity conditions, assuming that water supply is not a limiting factor.
Representative concentration pathway	A pathway of greenhouse gas emissions that leads to a particular concentration by the year 2100, adopted by the Intergovernmental Panel on Climate Change for its fifth assessment report (IPCC 2014). There are four pathways, expressed as the amount of extra radiative forcing in Wm^{-2} in 2100 produced by greenhouse gases: RCP2.6, RCP4.5, RCP6, and RCP8.5. Under RCP4.5, emissions are assumed to peak around 2040, then decline. Under RCP8.5, emissions are projected to continue to rise throughout the 21st century.
Resilience	A measure of the current ability of a species, community, or ecosystem to resist, absorb, and recover from the effects of hazards, by quickly preserving or restoring its essential basic structures, functions, and identity.

References

Abiad, A., R. M. Arao, S. Dagli, B. Ferrarini, I. Noy, P. Osewe, J. Pagaduan, D. Park, and R. Platitas. 2020. The Economic Impact of the COVID-19 Outbreak on Developing Asia. *ADB Briefs.* No. 128. Manila: Asian Development Bank.

Administration for Land Affairs, Geodesy and Cartography (ALAGaC), Mongolia. Land Use GIS 2015 Database. Ulaanbaatar. Unpublished.

Ahrenstorff, T. D., O. Jensen, B. C. Weidel, and B. Mendsaikhan. 2012. Abundance, Spatial Distribution, and Diet of Endangered Hovsgol Grayling (*Thymallus nigrescens*). *Environmental Biology of Fishes.* 94 (2). pp. 465–476.

Angerer, J., G. Han, I. Fujisaki, and K. M. Havstad. 2008. Climate Change and Ecosystems of Asia with Emphasis on Inner Mongolia and Mongolia. *Rangelands.* 30 (3). pp. 46–51.

Asian Development Bank (ADB). 2015. *Report and Recommendation of the President to the Board of Directors: Proposed Administration of Grant to Mongolia for the Integrated Livelihoods Improvement and Sustainable Tourism in Khuvsgul Lake National Park Project.* Manila.

—————. 2017. *Country Partnership Strategy: Mongolia, 2017–2020—Sustaining Inclusive Growth in a Period of Economic Difficulty.* Manila.

—————. 2019. *Report and Recommendation of the President to the Board of Directors: Proposed Loans to Mongolia for the Sustainable Tourism Development Project.* Manila.

Batima, P. 2006. *Climate Change Vulnerability and Adaptation in the Livestock Sector of Mongolia.* Final report submitted for the Assessments of Impacts and Adaptations to Climate Change (AIACC) Project. No. AS 06. Washington, DC: International START Secretariat.

Batima, P., L. Natsagdorj, P. Gombluudev, and B. Erdenetsetseg. 2005. Observed Climate Change in Mongolia. *Assessments of Impacts and Adaptations to Climate Change Working Paper Series.* No. 12. Washington, DC: International START Secretariat.

Batsukh, N., and A. Belokurov. 2005. *Management Effectiveness Assessment of the Mongolian Protected Areas System using WWF's RAPPAM Methodology.* Gland, Switzerland: World Wide Fund for Nature (WWF).

Bayarjargal, E., D. Dashkhuu, R. Mijiddorj, M. Russel, and P. Singh. 2019. Impact of Climate on the NDVI of Northern Mongolia. *Journal of the Indian Society of Remote Sensing.* 48 (2). pp. 333–340.

Bezuijen, M. R. 2019. Blue Skies and Green Steppe—Developing Sustainable Tourism in Mongolia. Asian Development Blog. 4 March.

Chik, W. I., L. Zhu, L. Fan, T. Yi, G.-Y. Zhu, X.-J. Gou, Y. Tang, J. Xu, W.-P. Yeung, Z.-Z. Zhao, Z.-L. Yu, and H. Chen. 2015. *Saussurea involucrata:* A Review of the Botany, Phytochemistry and Ethnopharmacology of a Rare Traditional Herbal Medicine. *Journal of Ethnopharmacology.* 172. pp. 44–60.

Chimed-Ochir, B., T. Hertzman, N. Batsaikhan, D. Batbold, D. Sanjmyatav, Y. Onon, and B. Munkhchuluun. 2010. *Filling the Gaps to Protect the Biodiversity of Mongolia.* Ulaanbaatar: WWF Mongolia.

Chu, T., and X. Guo. 2012. Characterizing Vegetation Response to Climatic Variations in Hovsgol, Mongolia Using Remotely Sensed Time Series Data. *Earth Science Research.* 1 (2). pp. 279–290.

Clark, E. L., J. Munkhbat, S. Dulamtseren, J. E. M. Baillie, N. Batsaikhan, R. Samiya, and M. Stubbe, eds. 2006. *Mongolian Red List of Mammals.* Regional Red List Series. Volumes 1 and 2. London: Zoological Society of London.

Cooke, G. D. 2007. History of Eutrophic Lake Rehabilitation in North America with Arguments for Including Social Sciences in the Paradigm. *Lake and Reservoir Management*. 23 (4). pp. 323–329.

Dagvadorj, D., Z. Batjargal, and L. Natsagdorj, eds. 2014. *Mongolia Second Assessment Report on Climate Change 2014*. Ulaanbaatar: Ministry of Environment and Green Development of Mongolia.

Dashkhuu, D., J. P. Kim, J. A. Chun, and W.-S. Lee. 2015. Long-Term Trends in Daily Temperature Extremes over Mongolia. *Weather and Climate Extremes*. 8. pp. 26–33.

Davi, N., G. Jacoby, K. Fang, J. Li, R. D'Arrigo, N. Baatarbileg, and D. Robinson. 2010. Reconstructing Drought Variability for Mongolia Based on a Large-Scale Tree Ring Network: 1520–1993. *Journal of Geophysical Research: Atmospheres*. 115 (D22).

De Grandpré, L., J. C. Tardif, A. E. Hessl, and N. Pederson. 2011. Seasonal Shift in the Climate Responses of *Pinus sibirica, Pinus sylvestris*, and *Larix sibirica* Trees from Semi-Arid, North-Central Mongolia. *Canadian Journal of Forest Research*. 41 (6). pp. 1242–1255.

ecoLeap Foundation. Khovsgol Litter Project. Hovsgol National Park Litter Education and Control Project (accessed 26 May 2020).

Elsen, P. R., W. B. Monahan, E. R. Dougherty, and A. M. Merenlender. 2020. Keeping Pace with Climate Change in Global Terrestrial Protected Areas. *Science Advances*. 6 (25). eaay0814.

Enkhtaivan, D. 2014. Modern Conditions and Recreational Loads on the Natural Complexes of the Khuvsgul Lake National Park. In Conference Proceedings of the Recreational Geography and Innovations in Tourism. Irkutsk. 22–25 September. Irkutsk: Publishing House of the V.B. Sochavy Institute of Geography, SB RAS. pp. 191–193 (in Russian).

Enkhtaivan, D., and O. V. Evropropyava. 2015. Transborder Tourism in Mongolia: Modern Problems of Service and Tourism. *Scientific and Practical Journal*. 9 (4). pp. 37–42.

Enkhtaivan, D., and B. Munguntulga. 2019. *Re-zoning of Khuvsgul Lake National Park*. Consultant's final report. Manila: ADB (TA 9183-MON).

Everard, M., P. Johnston, D. Santillo, and C. Staddon. 2020. The Role of Ecosystems in Mitigation and Management of Covid-19 and Other Zoonoses. *Environmental Science and Policy*. 111. pp. 7–17.

Fan, F., X. Dong, X. Fang, F. Xue, F. Zheng, and J. Zhu. 2017. Revisiting the Relationship between the South Asian Summer Monsoon Drought and El Niño Warming Pattern. *Atmospheric Science Letters*. 18 (4). pp. 175–182.

Fan, M. 2020. Achieving Sustainable Integrated Water Resources Management in Mongolia: The Role of River Basin Organizations. *ADB Briefs*. No. 138. Manila: ADB.

Foden, W. B., S. H. M. Butchart, S. N. Stuart, J.-C. Vié, H. R. Akçakaya, A. Angulo, L. M. DeVantier, A. Gutsche, E. Turak, L. Cao, S. D. Donner, V. Katariya, R. Bernard, R. A. Holland, A. F. Hughes, S. E. O'Hanlon, S. T. Garnett, C. H. Sekercioglu, and G. M. Mace. 2013. Identifying the World's Most Climate Change Vulnerable Species: A Systematic Trait-Based Assessment of All Birds, Amphibians and Corals. *PLoS ONE*. 8 (6). e65427.

Free, C. M., O. P. Jensen, S. A. Mason, M. Eriksen, N. J. Williamson, and B. Boldgiv. 2014. High-Levels of Microplastic Pollution in a Large, Remote, Mountain Lake. *Marine Pollution Bulletin*. 85 (1). pp. 156–163.

Free, C. M., O. P. Jensen, and B. Mendsaikhan. 2016. A Mixed-Method Approach for Quantifying Illegal Fishing and Its Impact on an Endangered Fish Species. *PLoS ONE*. 11 (1). e0148007.

Geyer, J., S. Kreft, F. Jeltsch, and P. L. Ibisch. 2017. Assessing Climate Change-Robustness of Protected Area Management Plans—The Case of Germany. *PLoS ONE*. 12 (10). e0185972.

Gilbert, R. O. 1987. *Statistical Methods for Environmental Pollution Monitoring*. New York: Van Nostrand Reinhold Company Inc.

Goulden, C., and M. N. Goulden. 2013. Adaptation to a Changing Climate in Northern Mongolia. In C. R. Goldman, M. Kumagai, and R. D. Robarts, eds. *Climatic Change and Global Warming of Inland Waters: Impacts and Mitigation for Ecosystems and Societies.* Oxford: John Wiley & Sons, Ltd. pp. 385–394.

Goulden, C. and R. McIntosh. 2018. The Critical Importance of Science and the Rule of Law in Protecting the Integrity of Mongolia's Lake Hövsgöl National Park. In Proceedings of the 12th Annual International Mongolian Studies Conference. Washington, DC. 9–10 February. pp. 1–16.

Goulden, C. E., J. Mead, R. Horwitz, M. Goulden, B. Nandintsetseg, S. McCormick, B. Boldgiv, and P. S. Petraitis. 2016. Interviews of Mongolian Herders and High Resolution Precipitation Data Reveal an Increase in Short Heavy Rains and Thunderstorm Activity in Semi-Arid Mongolia. *Climatic Change.* 136 (2). pp. 281–295.

Goulden, C. E., T. Sitnikova, J. Gelhaus, and B. Boldgiv, eds. 2006. *The Geology, Biodiversity and Ecology of Lake Hövsgöl (Mongolia).* Leiden: Backhuys Publishers.

Goulden, C. E., J. Tsogtbaatar, S. Chuluunkhuyag, W. C. Hession, D. Tumurbaatar, C. Dugarjav, C. Cianfrani, P. Brusilovskiy, G. Namkhaijantsen, and R. Baatar. 2000. The Mongolian LTER: Hovsgol National Park. *Korean Journal of Ecology.* 23. pp. 135–140.

Government of Mongolia. 2015. *National Biodiversity Programme, 2015–2025.* Ulaanbaatar.

―――. 2017. *Road Development Programme, 2017–2021.* Ulaanbaatar.

―――. 2020. *Vision 2050.* Ulaanbaatar.

Gradel, A., C. Haensch, B. Ganbaatar, and B. Dovdondemberel. 2017. Response of White Birch (*Betula platyphylla* Sukaczev) to Temperature and Precipitation in the Mountain Forest Steppe and Taiga of Northern Mongolia. *Dendrochronologia.* 41. pp. 24–33.

Hayami, Y., M. Kumagai, M. Maruo, T. Sekino, S. Tsujimura, and J. Urabe. 2006. Review of Some Physical Processes in Lake Hövsgöl. In C. E. Goulden, T. Sitnikova, J. Gelhaus, and B. Boldgiv, eds. *The Geology, Biodiversity and Ecology of Lake Hövsgöl (Mongolia).* Leiden: Backhuys Publishers. pp. 115–124.

Heiner, M., D. Galbadrakh, N. Batsaikhan, Y. Bayarjargal, J. Keisecker, O. Enkhtuya, O. Binderya, B. Tsogtsaikhan, D. Dash, B. Oyungerel, G. Purevbaatar, D. Zuberelmaa, and M. Munkhzul. 2017. *Identifying Conservation Priorities in the Face of Future Development: Applying Development by Design in the Khangai and Khuvsgul.* Ulaanbaatar: The Nature Conservancy Mongolia Program.

Heiner, M., D. Galbadrakh, N. Batsaikhan, B. Yunden, J. Oakleaf, T. Battsengel, J. Evans, and J. Kiesecker. 2019. Making Space: Putting Landscape-Level Mitigation into Practice in Mongolia. *Conservation Science and Practice.* 1 (7). e110.

Hessl, A. E., P. Brown, O. Byambasuren, S. Cockrell, C. Leland, E. Cook, B. Nachin, N. Pederson, T. Saladyga, and B. Suran. 2016. Fire and Climate in Mongolia (1532–2010 Common Era). *Geophysical Research Letters.* 43 (12). pp. 6519–6527.

Hijmans, R. J. 2015. FutureClim: 30-Seconds Downscaled Global Climate Model (GCM) Data from CMIP5. Available at WorldClim.

Hijmans, R. J., S. E. Cameron, J. L. Parra, P. G. Jones, and A. Jarvis. 2005. Very High Resolution Interpolated Climate Surfaces for Global Land Areas. *International Journal of Climatology.* 25 (15). pp. 1965–1978.

Hole, D. G., S. G. Willis, D. J. Pain, L. D. Fishpool, S. H. M. Butchart, Y. C. Collingham, C. Rahbek, and B. Huntley. 2009. Projected Impacts of Climate Change on a Continent-Wide Protected Area Network. *Ecology Letters.* 12 (5). pp. 420–431.

Intergovernmental Panel on Climate Change (IPCC). 2014. *Climate Change 2014: AR5 Synthesis Report.* Geneva.

International Permafrost Association (IPA). 2015. What is Permafrost?

International Union for Conservation of Nature (IUCN). 2008. *World Heritage Nomination–IUCN Technical Evaluation: Hovsgol Lake and Its Watershed (Mongolia).* ID No. 1082. Geneva.

————. 2020. *IUCN Green List of Protected and Conserved Areas.*

Japan Aerospace Exploration Agency (JAXA)/Earth Observation Research Center (EORC). 2020. Advanced Land Observing Satellite (ALOS) Global Digital Surface Model: ALOS World 3D – 30m (AW3D30).

Jeppesen, E., B. Kronvang, T. B. Jørgensen, S. E. Larsen, H. E. Andersen, M. Søndergaard, L. Liboriussen, R. Bjerring, L. Johansson, D. Trolle, and T. L. Lauridsen. 2013. Recent Climate-Induced Changes in Freshwaters in Denmark. In C. R. Goldman, M. Kumagai, and R. D. Robarts, eds. *Climatic Change and Global Warming of Inland Waters: Impacts and Mitigation for Ecosystems and Societies.* Oxford: John Wiley & Sons, Ltd. pp. 155–172.

Kaus, A., S. Michalski, B. Hänfling, D. Karthe, D. Borchardt, and W. Durka. 2019. Fish Conservation in the Land of Steppe and Sky: Evolutionarily Significant Units of Threatened Salmonid Species in Mongolia Mirror Major River Basins. *Ecology and Evolution.* 9 (6). pp. 3416–3433.

Knutti, R., G. Abramowitz, M. Collins, V. Eyring, P. J. Gleckler, B. Hewitson, and L. Mearns. 2010. Good Practice Guidance Paper on Assessing and Combining Multi Model Climate Projections. In T. F. Stocker, Q. Dahe, G.-K. Plattner, M. Tignor, and P. Midgley, eds. *Meeting Report of the Intergovernmental Panel on Climate Change (IPCC) Expert Meeting on Assessing and Combining Multi Model Climate Projections.* IPCC Working Group I. Technical Support Unit. Bern, Switzerland: University of Bern.

Kozhova, O. M., O. Shagdarsuren, A. Dashdorzh, and N. Sodnom. 1989. *Atlas of Lake Hövsgöl.* Moscow: Cartographic Ministry of USSR (in Russian).

Liancourt, P., B. Boldgiv, D. S. Song, L. A. Spence, B. R. Helliker, P. S. Petraitis, and B. B. Casper. 2015. Leaf-Trait Plasticity and Species Vulnerability to Climate Change in a Mongolian Steppe. *Global Change Biology.* 21 (9). pp. 3489–3498.

Liu, Y. Y., J. P. Evans, M. F. McCabe, R. A. M. de Jeu, A. I. J. M. van Dijk, A. J. Dolman, and I. Saizen. 2013. Changing Climate and Overgrazing Are Decimating Mongolian Steppes. *PloS ONE.* 8 (2). e57599.

MacKinnon, J., X. Yan, I. Lysenko, S. Chape, I. May, and C. Brown, eds. 2005. *GIS Assessment of the Status of Protected Areas in East Asia.* Cambridge: United Nations Environment Programme–World Conservation Monitoring Centre/Gland, Switzerland: IUCN.

Mattioli. S. 2011. Family *Cervidae* (Deer). In D. E. Wilson and R. A. Mittermeier, eds. *Handbook of the Mammals of the World, Volume 2: Hoofed Mammals.* Barcelona: Lynx Edicions. pp. 350–443.

McIntosh, S. 2017. Transportation Infrastructure Planning in Hovsgol Aimag's National Parks, Mongolia. Bachelor of Science degree thesis. Unpublished.

Meehl, G. A., and S. Bony. 2011. Introduction to CMIP5. *Clivar Exchanges.* 16 (2). pp. 4–5.

Metzger, M. J., R. G. H. Bruce, R. H. G. Jongman, R. Sayre, A. Trabucco, and R. Zomer. 2013. A High-Resolution Bioclimate Map of the World: A Unifying Framework for Global Biodiversity Research and Monitoring. *Global Ecology and Biogeography.* 22 (5). pp. 630–638.

Ministry of Environment, Green Development and Tourism (MEGDT), Mongolia. 2014a. *Integrated Water Resource Management Plan for the Khuvsgul Lake and Eg River Basin.* Ulaanbaatar.

————. 2014b. *Khuvsgul Lake National Park Management Plan, 2015–2020.* Ulaanbaatar.

Ministry of Environment and Tourism (MET), Mongolia. 2018. *Mongolia Third National Communication.* Ulaanbaatar.

————. 2019. *National Protected Areas GIS Database.* Ulaanbaatar: Division of Cadastre on Forest, Water, and Protected Areas.

————. 2020. *Mongolia: Integrated Livelihoods Improvement and Sustainable Tourism in Khuvsgul Lake National Park Project.* Completion report. Ulaanbaatar.

Mongol Ecology Center (MEC). 2013. Lake Hovsgol National Park: General Management Plan 2013. Foundation document. Ulaanbaatar.

Munkhchuluun, B., and B. Chimeddorj. 2013. *Management Effectiveness of Protected Areas in Altai Sayan Ecoregion and Amur Heilong Ecoregion Complex, Mongolia.* Ulaanbaatar: WWF Mongolia.

Namkhaijantsan, G. 2006. Climate and Climate Change of the Hövsgöl Area. In C. E. Goulden, T. Sitnikova, J. Gelhaus, and B. Boldgiv, eds. *The Geology, Biodiversity and Ecology of Lake Hövsgöl (Mongolia).* Leiden: Backhuys Publishers. pp. 63–76.

Nandintsetseg, B., J. S. Greene, and C. E. Goulden. 2007. Trends in Extreme Daily Precipitation and Temperature near Lake Hövsgöl, Mongolia. *International Journal of Climatology.* 27 (3). pp. 341–347.

Nandintsetseg, B., and M. Shinoda. 2013. Assessment of Drought Frequency, Duration, and Severity and Its Impact on Pasture Production in Mongolia. *Natural Hazards.* 66 (2). pp. 995–1008.

National Statistical Office of Mongolia. 2019. General Statistical Database (accessed 6 June 2019).

Natural Sustainable. 2019. *Integrated Livelihoods Improvement and Sustainable Tourism in Khuvsgul Lake National Park Project: Water Quality Monitoring Programme* (prepared for ADB). Final consultant report (Grant 9183-MON). Ulaanbaatar.

Nergui, S. 2020. *Integrated Livelihoods Improvement and Sustainable Tourism in Khuvsgul Lake National Park Project: Community Managed Loan Revolving Fund's Operation and Its Sustainability* (prepared for ADB). Final consultant report (Grant 9183-MON). Ulaanbaatar.

Nyamjav, B., J. G. Goldammer, and H. Uibrig. 2007. The Forest Fire Situation in Mongolia. *International Forest Fire News.* 36 (January–July). pp. 46–66.

Ocock, J., G. Baasanjav, J. E. M. Baillie, M. Erdenebat, M. Kottelat, B. Mendsaikhan, and K. Smith, eds. 2006. *Mongolian Red List of Fishes.* London: Zoological Society of London.

Olson, K. W., T. J. Krabbenhoft, T. R. Hrabik, B. Mendsaikhan, and O. P. Jensen. 2019. Pelagic–Littoral Resource Polymorphism in Hovsgol Grayling *Thymallus nigrescens* from Lake Hovsgol, Mongolia. *Ecology of Freshwater Fish.* 28 (3). pp. 411–423.

Opdam, P., and D. Wascher. 2004. Climate Change Meets Habitat Fragmentation: Linking Landscape and Biogeographical Scale Levels in Research and Conservation. *Biological Conservation.* 117 (3). pp. 285–297.

Oyuungerel, B., and O. Munkhdulam. 2011. Present Status of Specially Protected Natural Territories of Mongolia. *Geography and Natural Resources.* 32 (2). pp. 190–194.

Parmesan, C., and G. Yohe. 2003. A Globally Coherent Fingerprint of Climate Change Impacts across Natural Systems. *Nature.* 421 (6918). pp. 37–42.

Safronov, G. P. 2006. Gammaridae of Lake Hövsgöl. In C. E. Goulden, T. Sitnikova, J. Gelhaus, and B. Boldgiv, eds. *The Geology, Biodiversity and Ecology of Lake Hövsgöl (Mongolia).* Leiden: Backhuys Publishers. pp. 233–252.

Sharkhuu, N. 2006. Geocryological Conditions of the Hövsgöl Mountain Region. In C. E. Goulden, T. Sitnikova, J. Gelhaus, and B. Boldgiv, eds. *The Geology, Biodiversity and Ecology of Lake Hövsgöl (Mongolia).* Leiden: Backhuys Publishers. pp. 49–61.

Sharkhuu, A., A. F. Plante, O. Enkhmandal, C. Gonneau, B. B. Casper, B. Boldgiv, and P. Petraitis. 2016. Soil and Ecosystem Respiration Responses to Grazing, Watering and Experimental Warming Chamber Treatments across Topographical Gradients in Northern Mongolia. *Geoderma.* 269. pp. 91–98.

Sharkhuu, A., N. Sharkhuu, B. Etzelmüller, E. S. Flo Heggem, F. E. Nelson, N. I. Shiklomanov, C. E. Goulden, and J. Brown. 2007. Permafrost Monitoring in the Hovsgol Mountain Region, Mongolia. *Journal of Geophysical Research: Earth Surface.* 112 (F2).

Shimaraev, M. N., and V. M. Domysheva. 2013. Trends in Hydrological and Hydrochemical Processes in Lake Baikal under Conditions of Modern Climate Change. In C. R. Goldman, M. Kumagai, and R. D. Robarts, eds. *Climatic Change and Global Warming of Inland Waters: Impacts and Mitigation for Ecosystems and Societies.* Oxford: John Wiley & Sons, Ltd. pp. 43–66.

Shvidenko, A. Z., and D. G. Schepaschenko. 2013. Climate Change and Wildfires in Russia. *Contemporary Problems of Ecology.* 6 (7). pp. 683–692.

Sideleva, V. G. 2006. Fish Fauna of Lake Hövsgöl and Selenga River in Comparison with Ichthyofauna of Lake Baikal. In C. E. Goulden, T. Sitnikova, J. Gelhaus, and B. Boldgiv, eds. *The Geology, Biodiversity and Ecology of Lake Hövsgöl (Mongolia).* Leiden: Backhuys Publishers. pp. 357–378.

Simonov, E., O. Goroshko, and T. Tkachuk. 2018. Daurian Steppe Wetlands of the Amur-Heilong River Basin (Russia, China, and Mongolia). In C. M. Finlayson, G. R. Milton, C. Prentice, and N. C. Davidson, eds. *The Wetland Book.* Dordrecht: Springer. pp. 1499–1508.

Singh, N. J., and E. J. Milner-Gulland. 2011. Conserving a Moving Target: Planning Protection for a Migratory Species as Its Distribution Changes. *Journal of Applied Ecology.* 48 (1). pp. 35–46.

Sitnikova, T., C. Goulden, and D. Robinson. 2006. On Gastropod Mollusks from Lake Hövsgöl. In C. E. Goulden, T. Sitnikova, J. Gelhaus, and B. Boldgiv, eds. *The Geology, Biodiversity and Ecology of Lake Hövsgöl (Mongolia).* Leiden: Backhuys Publishers. pp. 233–252.

Spence, L. A., P. Liancourt, B. Boldgiv, P. S. Petraitis, and B. B. Casper. 2014. Climate Change and Grazing Interact to Alter Flowering Patterns in the Mongolian Steppe. *Oecologia.* 175 (1). pp. 251–260.

Spoelder, P. and Z. Batjargal. 2013. *Feasibility of Concessions for Tourism and Ecosystem Services in Mongolia's Protected Areas.* Final report for the Strengthening of Protected Area Network in Mongolia Project. Ulaanbaatar: United Nations Development Programme in Mongolia.

Stolton, S., N. Dudley, A. Belokurov, M. Deguignet, N. D. Burgess, M. Hockings, F. Leverington, K. MacKinnon, and L. Young. 2019. Lessons Learned from 18 Years of Implementing the Management Effectiveness Tracking Tool (METT): A Perspective from the METT Developers and Implementers. *PARKS.* 25 (2). pp. 79–92.

Terbish, Kh., E. L. Clark, J. E. M. Baillie, and J. Munkhbat. 2007. Proceedings of the Second International Mongolian Biodiversity Databank Workshop: Assessing the Conservation Status of Mongolian Reptiles and Amphibians. *Mongolian Journal of Biological Sciences.* 5 (1–2). pp. 19–28.

Tian, F., U. Herzschuh, S. Mischke, and F. Schlütz. 2014. What Drives the Recent Intensified Vegetation Degradation in Mongolia—Climate Change or Human Activity? *The Holocene.* 24 (10). pp. 1206–1215.

Triepke, F. J., B. McIntosh, O. Batkhuu, and B. Boldgiv. 2013. Natural Resource Management Planning Lake Hövsgöl National Park, Mongolia. United States Department of Agriculture (USDA) Forest Service Trip Report. USDA Forest Service. Unpublished.

Tsogtsaikhan, P., B. Mendsaikhan, G. Jargalmaa, B. Ganzorig, B. C. Weidel, C. M. Filosa, C. M. Free, T. Young, and O. P. Jensen. 2017. Age and Growth Comparisons of Hovsgol Grayling (*Thymallus nigrescens* Dorogostaisky, 1923), Baikal Grayling (*T. baicalensis* Dybowski, 1874), and Lenok (*Brachymystax lenok* Pallas, 1773) in Lentic and Lotic Habitats of Northern Mongolia. *Journal of Applied Ichthyology.* 33 (1). pp. 108–115.

United Nations Environment Programme–World Conservation Monitoring Centre (UNEP–WCMC), IUCN, and National Geographic Society (NGS). 2018. *Protected Planet Report 2018.* Cambridge, UK: UNEP-WCMC/ Gland, Switzerland: IUCN/Washington, DC: NGS.

Urabe, J., T. Sekino, Y. Hayami, M. Maruo, S. Tsujimura, M. Kumagai, B. Boldgiv, and C. E. Goulden. 2006. Some Biological and Chemical Characteristics of Lake Hövsgöl. In C. E. Goulden, T. Sitnikova, J. Gelhaus, and B. Boldgiv, eds. *The Geology, Biodiversity and Ecology of Lake Hövsgöl (Mongolia).* Leiden: Backhuys Publishers. pp. 387–402.

Vandandorj, S., E. Munkhjargal, B. Boldgiv, and B. Gantsetseg. 2017. Changes in Event Number and Duration of Rain Types over Mongolia from 1981 to 2014. *Environmental Earth Sciences.* 76 (2). p. 70.

Water, Sanitation, and Hygiene (WaSH) Action of Mongolia. 2020. *Integrated Livelihoods Improvement and Sustainable Tourism in Khuvsgul Lake National Park Project: Waste Management* (prepared for ADB). Final consultant report (Grant 9183-MON). Ulaanbaatar.

Watson, J., M. Rao, K. Ai-Li, and X. Yan. 2012. Climate Change Adaptation Planning for Biodiversity Conservation: A Review. *Advances in Climate Change Research.* 3 (25). pp. 1–11.

Whaller, P. 2015. Terminal Evaluation: United Nations Development Programme/Global Environment Facility (GEF) Integrated Natural Resource Management in the Baikal Basin Transboundary Ecosystem. Mongolia and Russian Federation. Washington, DC: GEF.

World Health Organization (WHO). 2020. WHO Coronavirus Disease (COVID-19) Dashboard (accessed 18 July 2020).

WWF Mongolia Programme Office. 2011. *Assessments of Climate Change and Anthropogenic Impacts into Hydrological Systems of Onon, Kherlen and Khalkh River Basins, Mongolia.* Ulaanbaatar.

Yu, L., and M. Goulden. 2006. A Comparative Analysis of International Tourists' Satisfaction in Mongolia. *Tourism Management.* 27 (6). pp. 1331–1342.

Zhang, Y., H. Enomoto, T. Ohata, and H. Kitabata. 2017. Glacier Mass Balance and Its Potential Impacts in the Altai Mountains over the Period 1990–2011. *Journal of Hydrology.* 553. pp. 662–677.

Zomer, R. J., A. Trabucco, D. A. Bossio, and L. Verchot. 2008. Climate Change Mitigation: A Spatial Analysis of Global Land Suitability for Clean Development Mechanism Afforestation and Reforestation. *Agriculture, Ecosystems & Environment.* 126 (1–2). pp. 67–80.

Zomer, R. J., A. Trabucco, M. J. Metzger, M. Wang, K. P. Oli, and J. Xu. 2014. Projected Climate Change Impacts on Spatial Distribution of Bioclimatic Zones and Ecoregions within the Kailash Sacred Landscape of China, India, Nepal. *Climatic Change.* 125 (3–4). pp. 445–460.